I0203749

LOST MINES
OF CALIFORNIA
AND THE SOUTHWEST

— BY —

R. W. McALLISTER

COACHWHIP PUBLICATIONS
Greenville, Ohio

Lost Mines of California and the Southwest, by R. W. McAllister
First published 1953.
© 2014 Coachwhip Publications
No claim made on public domain material.
Cover: Mono Lake, California © Frosti1977.

ISBN 1-61646-228-0
ISBN-13 978-1-61646-228-4

CoachwhipBooks.com

ACKNOWLEDGMENTS

The information for the maps and articles has been gathered from prospectors and miners, from the State of California oil and mineral maps, journals of mines and geology, bulletin 141 (The mother lode country) mineral bulletins of California, U.S. Geological Survey Bulletin 507, from the Desert Magazine, National Geographic Magazine, the Arizona Highway, J. Frank Dobie's "Coronado's Children," from True Magazine, Josiah Gregg's Commerce of the Prairies, Philip Bailey's "Golden Mirages," John D. Mitchell, the Los Angeles Public Library and from old books some of which are out of print.

PREFACE

Herein an attempt has been made to correlate and condense more than two hundred historical tales and legends of lost mines. A supplementary map accompanies the articles and shows the locations about which the tales have been written. The articles are assigned numbers and letters corresponding to those on the map. In some instances little more than a place is noted, in others, a short condensed version of the tale so that it is only necessary for the adventurous nomad to furnish a little imagination as to the action which took place. Much of the buried treasure remains hidden and the mountains will continue to reward the seeker of gold. There is romance and adventure in the quest. Many of the best mines will be found. Recently, promising gold mines were found in the Chocolate Mountains and in the Cargo Muchacho Mountains—possibly, The Lost Padre and The Lost Pegleg mines.

(A) "Cabeza de Vaca" — 1522

Thirty years after Columbus landed on American soil, the Spaniards started the conquests and explorations which were to yield them considerable treasure. Pizarro conquered the Incas and obtained millions in gold and silver from their treasure houses. Perhaps the land to the north would yield greater treasures. Narvaez had explored Florida and was told of wealth to the North and West. Cabeza de Vaca, with three score adventurers, followed Narvaez and had his boats wrecked on the shores of Louisiana. Only a few men and de Vaca survived. They searched for gold to the north, heard many tales from the Indians which fired their imaginations and were finally made captives by the Indians and relegated to the status of slaves. After several years they escaped and began the long walk across New Mexico to the Pacific, being the first transcontinental travelers. During the eight years of their homeward journey, stories continued to come to them of great cities to the north with houses studded with gold. When they arrived in Mexico City, they repeated the stories they had heard.

Friar Marcos de Niza was commissioned by the viceroy to check on de Vaca's stories. Estevan, the negro who had been with de Vaca, acted as guide. He traveled ahead of the expedition, and before he was killed by Indians, sent back word of marvelous cities and great things to come. Friar Marcos returned to Mexico City to recruit a larger expedition.

(B) "Francisco Vasquez de Coronado" 1540

In February of 1540, Francisco Vasquez de Coronado was put in charge of the expedition. His command was composed of 250 cavalrymen, 200 foot soldiers, 1000 Indians, 1000 horses and mules, pack trains and great herds of cattle, sheep, goats and pigs. Coronado, with some picked men, moved on ahead of the cumbersome army. He traveled northward as far as the present city of Wichita, Kansas, in search of Gran Quivira. His headquarters for two winters was at Tiguex, an Indian pueblo near the present town of Bernalillo, New Mexico. The quest for Gran Quivira with its treasure houses of gold, failed, but one of the greatest land explorations was accomplished.

(C) "Don Antonio de Espejo" — 1582

In 1582, Don Antonio de Espejo, with a force of fifteen men, explored around Zia and Jemez, past Acoma and Zuni and west to Prescott, Arizona, in search for a great lake on whose shores were many rich settlements. Then he explored to the east as far as the Pecos River, which he followed southerly for 300 miles, returning to San Bartolome, Mexico.

(D) "Juan de Onate" — 1596

Juan de Onate's ambition was to explore to the Pacific. He financed his project, including 200 soldiers, 83 wagons, 400 men, women and children, eight Franciscan padres, four lay brothers, and 7000 head of livestock. He visited the Tiguex Province, Zia, San Juan, Picuris, Taos, Pecos and Santo Domingo. He discovered the salt lakes of the Abo region. Westward he reached the Hopi country, but at Acoma, part of his band was destroyed and he gave up the idea of exploring to the Pacific. Instead, he turned eastward and saw the plains of Texas and Kansas.

(1) "Aztec Treasure" — 1611 Mexico

The Spaniards learned that the Aztecs possessed large amounts of gold, silver and turquoise. An expedition was sent from San Miguel de Culiacan on the coast of Nueva Viscaya and proceeded to the Intendencia de Sonora, State of Sinaloa. Black Stephen, the Moor, acted as guide. As was customary on the approach of the Spaniards, the Indians buried the treasure in a cave which was located at the edge of a valley six leagues long. The Indians were defeated, but Black Stephen was carried off and lived in captivity with the Indians, dying at the age of 97. A family at Choix, Mexico, had the record. (Look up members of the Don Inocenti Cinfuentes.)

(2) "Mesa of the Bulls" — 1611 Mexico

Estevan, the negro mozo who was with Cabeza de Vaca, made a record on Tepolcates, which is still in existence. From his record it is necessary to find the Mesa of the Bulls, or Mesa del Toro, thirty miles from Choix on the Choix River. Indian scouts had reported the approach of the Spaniards. Their king ordered all treasure brought to him and described the Cueva Pintada (painted cave) thusly: "I stand with my back to Toro (large rock resembling a bull) and where the shadow strikes on the opposite bank of the arroyo in the morning, all valuables must be placed in a cave built for the purpose." The treasure amounts to several tons of gold and silver ornaments. As of this day, placer miners wash out small ornaments in the Arroyo del Toro. Arrowheads, tomahawks and rock crosses abound.

(3) "Treasure at Bachaca" — 1769 Mexico

Spanish miners worked on a very rich silver vein at the Bachaca (Waterfall), which lies seven leagues to the southwest of Alamos. Water from two arroyos enters the valley, unite and pass through a narrow gorge over a shelf of rock into a shady pool. A dam was placed across the narrow outlet. From a distance the Pacific Ocean can be seen, and the Mayo River is just to the west. Considerable treasure was taken out between the years 1769 and 1823. Then a large band of Chihuahua Indians came through Chinipas Pass and wiped out the miners then working and attacked the Alamos. Years later,

when excavating near Bachaca, workmen uncovered a concrete slab with an iron ring. It covered a cellar which contained three large ollas filled with Spanish money and the room contained $100,000.00 worth of silver bullion.

(E) "Escalante and Dominguez" 1776

In 1776 Franciscan friars in New Mexico desired a safer route to the Pacific than the one through Zuni and the Apache country in Arizona. It would open up a new trade route if they could get through by way of the country to the north and would also join the mission of Monterey in California, and Santa Fe in New Mexico. Father Silvestre Valez de Escalante and Father Francisco Anastasio Dominguez left Santa Fe to blaze the new route to the coast. They followed the course of the Chama River to the Colorado line and thence northwestward until they reached Utah Lake, just south of the present Salt Lake City. The numerous hardships took their toll; with winter coming on and no one available who was familiar with a route from Utah Lake to the west, they decided to alter their course and return to Santa Fe. After much difficulty, they negotiated a crossing of the Colorado River at what was called the "Crossing of the Fathers"—now Lee's Ferry.

(F) 1786

Pedro Vial opened a trade route between San Antonio, Texas, and Santa Fe, using old Indian trails. It took him a year to make the trip. Then in 1790, he made the hazardous trip to St. Louis to survey that route for trade purposes.

In a very short time after these trade routes were established, commerce picked up and goods of considerable value passed over these old Comanche trails. It was dangerous business, however, and at any time the Indians of the prairies were apt to swoop down on the caravans, killing, plundering or making it necessary to bury the treasures.

(G) "Zebulon Pike" — 1805

With Zebulon Pike, James Burcell and Captain William Becknell pioneered the trail of commerce known as the Santa Fe Trail. The Spaniards and Indians discouraged trade from the East but after 1822, when Mexico won her independence from Spain, long caravans wound across the plains from Independence over the Raton Pass. One firm alone had 3500 wagons, 4500 men, 40,000 oxen and 1000 mules.

Then followed home seekers with their lumbering prairie schooners and stage coaches loaded with passengers headed west to seek their fortunes.

(H) 1839

Josiah Gregg and party surveyed a southern and northern wagon route into Santa Fe. The route across Oklahoma was mostly East to West. They carried guns, one being a long brass piece of 1¼ inch calibre that would throw a lead ball fully one mile accurately, much to the discomfiture and surprise of small raiding bands of Indians who hove into sight. The personnel multiplied by pairs and bands, as the group proceeded westward, being recruited from what remained of small bands of white men that had been attacked by Indians, by deserters from the Army, renegades and traders. Safety lay in greater numbers. Captain Sublette had a party of eighty with which to protect travelers. His band was attacked near Rabbit Ear Mountain about the time Gregg reached there, Captain Smith being among those killed.

Some members of Gregg's motley crew were hard to control. At the river crossings where grass was lush, and where grapes, wild plums and currants grew profusely, it was only natural that deer, turkey, partridge and grouse would be seen in abundance. And on the plains where the grass was thick and high, great herds of buffalo ranged. Gregg's hunters were told to bring in only enough game to supply the camp, but with excitement and enthusiasm high during the chase, much of the game was slaughtered to no purpose. Indians discovering such wanton killing over their hunting ground usually reacted fiercely, which made progress hard for the pioneers.

(J) 1849

The start of the great migration to the West came with the discovery of gold in California. New stage and freight routes were established. It was during this period when the longest route of all was marked out—the Butterfield Trail—reaching from Tipton, Missouri, to San Francisco, California. Passengers paid $100 in gold to travel eastward and $200 westward. The Apaches, under Victorio, staged many a raid.

Cook's Canyon was known as the "Gauntlet of Death" where 400 soldiers and emigrants lost their lives in spite of its being almost within hearing distance of old Fort Cummings. The traveler gave thanks when he passed Cooks Canyon safely, and was glad to leave lonesome "Soldier's Farewell." Nearing Arizona, he passed "Shakespeare," "Stein's Stand," "Doubtful Canyon," and soon entered San Simon Valley. After leaving the Gila River and Yuma, the next watering place was Carrizo. From Carrizo to Vallecito is sixty miles. Because of drifting sand and lack of water, this portion of the trail was abandoned and a route a little to the West was chosen into Warner's Ranch.

(K) "The Sonora Trail" (Camino del Diablo) — 1849

This old route from Sonora, Mexico, to Yuma, Arizona, took considerable toll. There was one watering place in 250 miles at the Yuma end. From Yuma to Carrizo Spring is 100 miles. From Tinajas Atlas to Sonoita is 150 miles, with no water except for 'tinajas' or mountain tanks. The entire crossing shows abandoned wagons, gun barrels, trunks, barrels, wearing apparel, bullets and powder canisters.

(4) "Two Suns East" — Arizona

Seventy-five miles, or two suns east of Ajo, on the northern edge of Santa Rosa County, is a rich gold ledge or chimney, running thousands of dollars to the ton.

An old Papago Indian was known in Casa Grande, Ajo and Gila Bend, where he traded and sold gold in quantity. Years after the old Indian died, a searching party was organized to look for the gold, plans having been made on the assumption that the grandson, with his inherited map, could lead them to it. Their route lay across Ajo Valley through dense mesquite and greasewood, and then followed the old Indian Trail, climbing to a large water tank near the summit of the first range. Here they found plenty of quail, doves and coyotes. In the center of the valley, beyond this summit, a tabletop mountain rises to a good height, the old Indian Village and burial ground being nearby. Two day's search failed to find the mine but a 35 pound piece of float in the Arroyo on the west side of the mountain panned $1200 in gold. Papagos insist the mine is located in a small—iron stained mountain having a stone slab over the entrance.

(5) "Lost Nigger Ben Mine"
1886 — Arizona

"Little Antelope" is one of the richest mines in Arizona, and is at Rich Hill in Yavapai County. This mine produced 20 millions in placer, and $780,000 in large nuggets which were picked up from the surface of Rich Hill. Gold amounting to $5000 or $6000 was often found under one large boulder. Mr. Peeples, an Arizona pioneer, took a fortune from the mine and started ranching in Peeples Valley. Nigger Ben, a ranch employee, was told by an Indian that there was a "Big Antelope" at Sycamore Spring. Ben was taken to this hill by the Indian twice, failed to return the second time, and was found dead by Peeples at Sycamore Springs. The gold of "Big Antelope" is yet to be found.

(6) "Lost Nugget Mine"
1900 — Arizona

About two days trip for a burro along the east side of Colorado River south of Topock, look for a streak of yellow clay full of gold nuggets. Legend tells us that two men took out $50,000 to $60,000, and that they were unable to go back to the mine.

A Santa Fe Railroad Conductor at Needles was shown about $30,000 worth of nuggets and was given some of them. This man and his son returned some months later, went into the desert south of Topock, but did not return. Searchers found them dead.

(7) "Caugh Oir — Golden Cup"
Arizona

Extinct geyser on top of Rich Hill about eight miles east of Congress Junction, Yavapai County, Arizona, at 5200' elevation, 2000' above surrounding plains. Believed to be the source of the $780,000 produced by Rich Hill. Crater now filled with granite boulders, iron-stained clay, pieces of iron and pebbles of silica. Some day, perhaps, tunnels will tap the mountains and release the nuggets therein.

(8) "Treasure of Don Felipe"
Mexico

The treasure is near the Town of Alamos— Rancho Santa Barbara, at the base of Guadulupe Mountain in the Sierra Madre Range. The Mayo River is in a narrow gorge, as it passes Guadalupe, makes a great bend around a high point of land that juts out to the North and it is here that Don Felipe's ranch house stood. The River from here flows through Mayo Valley to the Sea. Three times a week numbers of pack mules, loaded with gold and silver bars, came to Santa Barbara from the mines. There was silver on Guadulupe Mountain and gold at Sobia just over the mountains to the west. One-fifth of all the bullion went to Mexico City for the King of Spain.

Santa Barbara was attacked by Indians in 1810. Don Felipe was killed there and his daughter Clothilde was taken to Mexico City.

The gold mines at Sobia were relocated and produced large amounts of gold, but the Guadalupe silver mine has never been found.

Mayo Indians believe that buried gold and silver will glow in warm weather when the ground is wet. Every June, when the rains begin, the hunt goes on.

(9) "Lost Breyfogle Mine"
1863 — California

Breyfogle and companions, prospecting to the South of Las Vegas, Nevada, camped at a spring on the side of the high mountain. Three Indians took Breyfogle to a rich gold mine three miles away. He was left for dead by the Indians, after fracturing his skull, but he got back to camp and found his partner slain and the camp robbed. Making the trip back to the Vegas Ranch in three days, he left with Mrs. Stewart, owner, some of the ore which was a pale yellow carbonate full of black silver sulphides, horn silver, and was rich in gold. At Austin, Nevada, ore samples may be seen to this day. A camp, believed to be John Breyfogle's, was found in a canyon at a small spring in the McCullough Mountains, 14 miles northwest of the town of Searchlight. An old Mormon from Salt Lake City made several trips to the McCullough Mountains and returned with several burros loaded with rich ore. He was found shot to death on the bed of a dry lake on the west side of the mountains. Searchers called it the "LOST MORMON MINE."

(10) "Lost Sopon Mine" — Arizona

From Tucson, on the road to the old Spanish town of Arivaca in Southern Pine County, Arizona, stands the shrine of Santa Rita on a high promontory of the Sopon Ranch overlooking Sopon Creek, and comprises a cluster of old adobe buildings in cottonwoods. The Sopon Gold and Silver Mine produced rich ore worked with arrastres. The old mine was carelessly worked, the vein was lost and the gold produced was buried at the corner of the old adobe house. Apache Indians killed all white residents except a baby who grew to womanhood in Tucson.

(11) "Lost Tayope Mine"
1635 — Mexico

Located in the Sahuaripa District of Mexico, near Vacatete (Cowteat) Mountain, in the Yaqui Indian Country, State of Sonora, there is supposed to be the richest gold mine on the American Continent. In 1767, the Tayope was sealed up and abandoned. The Camp and church were destroyed a little later by Indians. Millions in gold bullion was sealed in the mine.

A Yaqui Indian reports the location of the mine as north of the Cerro Azul (Blue Mountain) and that the mine was an Antigua, having been worked by the Spaniards, and that there should be seen the ruins of an old camp with arrastres and a canal twelve kilometers long which brought water for use in the arrastres. The canal ends near the ruins of an old church.

(12) "Lluvià de Oro" (Shower of Gold) and "Gloria Pan Mines" — 1893 Richest Mines of Mexico

A Sinaloa Indian talked in his sleep giving away his secret. He took the San Juan Fiesta patron to the Lluvia de Oro Mine and received in payment the meat of an old bull—no hide. He refused to show the Gloria Pan Mine which was discovered in 1750 and operated until 1767, because the bull's hide was withheld. Gold piled up and was stored in the Mine which amounted to several millions in gold bars. The Jesuits sealed the mine before being expelled by Charles III of Spain.

The old patron took out several million from the Lluvia.

(13) "Lost Escalante Mine" 1698 — Arizona

It is located somewhere in the Santa Catalina Mountains north of Tucson, Arizona, and called the Mine of the Iron Door. The records were left by Padre Escalante, who travelled all over Pimeria Alta to collect the church share. Calistro, an Opata Indian, living on the Santa Cruz River, near the ancient ruins of Tumacacori Mission, told of the Escalante Mine which was worked by the Spanish in the Sixteenth Century. Calistro's grandfather and himself, while hunting wood rats, rested at a large rock near a pool of water in the bed of Canada del Oro (Grand Canyon). The grandfather pointed up to the Santa Catalina Mountains and said that the mine there was the richest ever discovered in Pimeria Alta (now Santa Catalinas). It was on the western slope not far from a ventana (hole in rocks resembling a window). Hostile Indians killed all miners remaining at the mine during a nearby fiesta. The entrance to the mine, containing a strong room filled with gold bars, was destroyed.

(14) "Lost Soapmaker Mine" 1850 — Arizona

A party of Mexicans encamped at Tinaja Alta, 70 miles west of Ajo, were making a crossing of the Colorado near Yuma to take the old Butterfield stage route to California. The animals strayed off and a Mexican, while climbing for a view of the lost animals, discovered a gold vein two feet wide on the south side of a small round mountain. The richest part of the vein is a strawberry colored quartz showing wire and coarse gold. The vein was gradually being covered by drifting sand brought in by windstorms from the south. To the west lies Cabeza Prieta (Black Head Mountain). The vein, which is almost one-third of the distance up the mountain, ran northwest and southeast. The finder left the party and returned to Caborca, his starting point, in Sonora, Mexico, to get the help of a German soapmaker at Caborca. He, together with two Mexicans, built an Arrastre at a small spring to the south. They were shot by Papago Indians who filled the shaft. A "Doctor" Juan, a Papago, says a wind from the north will some day uncover the wonderful vein from which a million dollars in gold could quickly be taken. He was one of the three Indians who murdered the soapmaker and party.

(15) "Lost Six-Shooter Mine" 1883 — Arizona

The superintendent of the Planet Mine, Yuma County, while traveling between the town of Quartzsite and his camp, became lost in a desert sand storm. He took shelter at a quartz ledge which outcropped above the sand, and found that the quartz was full of free gold. The ore taken from the dead superintendent's pockets, who was found near the edge of the desert, assayed $25,000 to the ton. A note was also found explaining that he had left his belongings at the ledge, including his two six shooters and coat. His horse, half-dead, returned to the Planet Mine.

(16) "The Glory Hole" 1909 — Arizona

Three prospectors, Alger, Barker and Griffin, searched in the Harquavar Mountains, twelve miles northwest of Salome in northern Yuma County, Arizona. Alger had found rich float 20 years before and had been looking every year since. About ready to give up, the three men discovered the vein. Three sacks of ore taken to Phoenix assayed $338,510 per ton. Salome and Vicksbury boomed, and the city of Winchester sprang up about three miles south of the Glory Hole. The mine was leased years later and some high grade ore was taken, but the rich vein was lost.

(17) "The Golden Beans" 16th Century — Mexico

From the town of Camoa on the Mayo River, the old padre could look in a southeasterly direction into a tunnel on the side of a large mountain known as Oro Mani (Golden Peanuts). The padre discovered the ledge from which the nuggets were eroded and a large amount of gold taken. Search for it started 200 years later. At the Church at Camoa, a solid gold crown and gold dagger were excavated. The ledge was not found, although the Indians claimed to have seen the strange light that gold produces during the rainy season. Still undiscovered.

(18) "Black Gold" — California

Way off to the northeast of Indio, California, and 60 miles from water, is a mineralized cone or extinct volcano The gold is tarnished black like the rocks. The district affords no rain to speak of and no vegetation. The crater is in a valley below a rimrock mesa. Two men reached the crater, finding $10,000 in gold beside some skeletons and the floor of the crater was covered with millions of dollars worth of gold among the rocks and volcanic ashes. They returned with $65,000. They were never able to make a second trip. The white man purchased an orange grove at Redlands.

(19) "Good Medicine and Buried Gold at Ajo" — Arizona

Papago Indians worked the arroyos around copper mines at Ajo for gold placer. Gold was traded for supplies at Caborea in the Altar district.

Mexicans took the workings easily from the non-warlike Papagos. Apache Indians on their way from the Gulf to their stronghold in the Superstition Mountains to gather fruit from the Saguaro Cactus which grows in the Superstitions, attacked the Mexicans who left guns, supplies and gold buried in the Camp. The Sonoita River, which was 35 miles south of the mines, was the first stop. The Papagos, enemies of the Apaches, regained their ground after a terrible storm and whirlwind drove the Apaches away. The buried gold was never found.

(20) "The Copper Box" 1623 — Arizona

Jesuits were mining in the vicinity of the old mining mission of Tumacacori on the Santa Cruz River south of Tucson. In 1850, Americans found and sold old slag from just east of the mission.

Papago Indians tradition says the bullion was packed out to the Guadalupe Mine one league southwest of the mission, and placed in the mine tunnel. There were 2050 mule loads of silver from Huachuca and the Santa Rita Mountains, and 905 loads of gold and silver bullion from the mines of Tumacacori and Tascasa Mountains—in all it was about 45 million pesos.

Church fixtures from Altar, treasure maps and description of treasure, were placed in a copper box, to be taken to the Altar Mission to the South. The trail from Tumacacori led to the Guadalupe Mine, thence up through the pines to the summit of Tumascacori Mountains with the Santa Cruz Valley and the old white mission below. The trail led past the Tumaca-

cori Mine on the west side of the mountain and one and one-half leagues from the Guadalupe Mine. Here, a pack train headed in the opposite direction and was met. The padre reported conditions bad at the Mission Altar. The treasure was buried there. The copper box must be found for clues for the rest of the treasure.

(21) "Lost Mines and Buried Treasure of the Tumacacori Mission Tuboc in the Santa Cruz Valley" 1784 — Arizona

The distance from Nogales on the State Highway to Tucson is about 22 miles. The Spanish mined here after Father Kino and Jesuit assistants were active. Gold and silver lies buried, brought from the Huachuca Mountains. From the Guadalupe Mine to the Pure Conception Mine is three leagues, being near the old town of Santa Cruz. The mine is covered by an iron door with copper handles. Native silver was mined weighing 25 to 250 pounds in slabs. Halfway between is the Apata Mine, and from Apata it is a league to the mine called Tumacacori where all the treasure of the Saint Guadalupe was left. The San Pedro Mine is a league and a half from the temple; the Isabella is one league north of the San Pedro; and a trail from here reaches the Spring of San Roman where there is gold and silver.

(22) "Buried Gold of Casa Grande" Mexico

Upper Valley of Rio Del Norte and Rio Gila —built during Montezuma's time. Coronado passed here in 1540. Pima tradition goes as follows: A woman of surprising beauty resided in a green spot in the mountains. As tribute, she received grain, skins, etc., but gave no love in return. When drought came, she aided distress from her endless supplies. One day she fell asleep with her body exposed; a drop of rain fell on her abdomen and produced conception. A son was the issue, who became the founder of a new race, and built the Casa Grande. In 1520, when the Spaniards imprisoned Montezuma and demanded gold, runners were sent out and all people advised to bury their gold. This was done and a great treasure is still to be found in a cave in the Salt River Mountains near the Yaqui Indian Village of Guadalupe.

Note: At Monte Albon, State of Michoacan, Mexico, a $25,000,000 treasure was unearthed.

(23) "Red Rock Treasure" — Arizona

Northwest of Tucson-Silver Bell Road near the town of Red Rock. Gold and silver from nearby mines was worked into ornaments by the ancient people, according to Papago tradition. White men came to live with the Indians and bartered for gold and silver. After several years and numerous trips into the mountains, a large amount of gold and silver was accumulated. Hostile Indians from the north killed them all. Their treasure was buried. These men are believed to have been a party from Leif Ericson's Colony on the Coast of Maine. The Aztec legend accounts for a white man who visited them in 922. He made several trips to the Aztec Capital and disappeared in 935. He was known as Quetzalcoatl.

(24) "Maximilian's Gold" — Mexico

Gold assigned to Emperor Maximilian at Mexico City left the gold mines in Yaqui country, southern Sonora, in 1866. When they reached the Mayo River, El Rancho Santa Barbara, where they hoped to strike the trail that led through the Chinipas Pass, they were advised that a revolution had broken out in Mexico. The 18 muleloads of gold were buried about one kilometer from the crossing. Chati Almada and his macheteras killed all but two of the soldiers.

(25) "Lost Peg-Leg Mine" — California

Peg-leg and another man set out for Los Angeles from Yuma or elsewhere to stalk one of the numerous Indians who brought gold to Yuma. The Indians were always lost sight of near Cottonwood Springs, San Bernardino County. Peg-leg was picked up near Walker's Station on the Southern Pacific Railroad, dying of thirst. He died in a hospital in Los Angeles. On his person was a large black nugget of gold. Three little hills, the largest with a black summit and chalky base, are located somewhere between Walker's Station, Warner's Pass and north of Cottonwood Springs.

(26) "Brady Mines" — Arizona

Peter R. Brady was engaged in farming at Florence, Arizona, in 1869. Juan Gradillo, an Indian friend of the Papago Indians, gave Brady the rich float that made the Vekol silver mine. The Vekol and Ajo were both mined successfully by Brady, being two of the richest of Arizona mines. He out-guessed a band of 500 Mexicans who demanded the mine at Ajo by showing a fictitious force of men who used a connecting tunnel and made a circuitous route between two entrances, each man passing in review several times.

(27) "Lost Squaw Mine" — Arizona

Two Yuma bucks and an old squaw on the trail between Phoenix and Yuma picked up a piece of very rich gold ore. The old squaw was badly treated by some Mexicans when she couldn't locate the vein for them. She told Ed Schieffelin the location but he couldn't find the vein. Later, the Harquahales was discovered which produced many millions and which according to Ed, was the "LOST SQUAW MINE."

(28) "Legend of Dr. Thorne and Origin of Squaw Tea" — Arizona

Dr. Thorne was captured by the Apaches about 1850. While the tribe was in the White Mountains, they camped at a stream about 30 or 40 miles from the Black River. On the rock bottom Thorne discovered gold. From his point of discovery, he could see Sombrero Butte and Sierra Pintades and the trail crossed Salt River near its junction with White River. Because of blindness, after being released from the Indians, Dr. Thorne could not look for gold but told others of its location as near as he could. An epidemic broke out among the squaws. Thorne gained his release by curing them with tea from a desert growth.

(29) "Mines of New Mexico"

Antonio De Espejo visited the ancient pueblos of Acoma, Zuni and Moqui in 1582. About forty-five leagues away he found the rich silver ore. Large scale mining started in 1725. Before 1680, when the Spaniards were expelled, many mines were operated. These were covered up by the Indians because of cruel oppression.

(30) "Planchas de Plata" — Arizona

A world famous silver mine, was discovered about 1730 in the Magdelena district near Nogales. The name signifies plates or bars of silver. A silver nugget weighing 2700 pounds was found. One sheet weighed 149 arrobas, another 21. In the Guadalupe Mine, near the Tumacacori Mission, 45 million pesos in gold and silver bars were buried. The surrounding district was rich and to the north lies the mine Ojitas de San Roman. About five leagues away is the Arizona mine; and the Presidio Santa Cruz is twenty-five leagues away. To find this silver go to the Baboquivari or Sierra del Pajorico Mountains, close to the Temijalta Trail, and look for dividing veins in silvery ash.

(31) "Lost Yuma Mine" — Arizona

Thomas McLain or "Yuma" carried on a trade with the Apaches from his ranch in the Papago country west of Tucson, meeting the Indians on Pinai Mountain. The Indians showed him a marvelously rich gold mine near the meeting place—their source of gold for trading purposes. With General Walker, future agent of the Papagos, who accompanied him in 1860, his last trip, they uncovered the loose earth over the ledge, broke off quartz with their hatchets, and hurried over to Tucson. The ore was so studded with gold that a pinhead could not touch the quartz between gold beads. Charles O. Brown, Grant, and Wm. S. Oury, Hy S. Stevens and Sam Hughes and A. Lazard vouched for the ore. Yuma was killed by the Papagos, who feared he was planning a raid on them with the Apaches. General Walker, who died in 1865, gave a description to John Sweeney, who, in turn, told Chas. O. Brown. The mine must be near the junction of the Gila and San Pedro Rivers.

(32) "Lost Squaw Hollow Mine" 1864 — Arizona

From Camp Creek as a base and Squaw Hollow as a source, many have searched for gold-bearing quartz. Soldiers under Colonel Woolsey—Judge J. T. Alsap being one of them, after defeating a band of Apaches, prospected near their camp and found the rich quartz. A sheep herder in the 80's reported a man, an outlaw, mining in Squaw Hollow. The operation was carried on in a wooded basin near a small cabin. The ore was crushed and panned at the creek alongside the cabin.

(33) "Lost Adams Diggings" 1882 — Arizona

Robert T. Emmet, 2nd Lieutenant, 9th Con., U.S.A., left a letter telling of rich placer. Seven men, including Adams, stopped at Camp Apache for rations. A few days east they found rich placer in a canyon. A German, becoming alarmed about the Indians, left for Fort Yuma and cashed in about twelve thousand dollars in gold. When supplies got low at the cabin in the canyon, all, but Adams and one other man, started for Fort Yuma. After a long delay, Adams and his companion started out to look for them and after going to a mountain nearby, glanced back to see their cabin in flames and their companions who had returned from the other direction, being massacred. Adams escaped and was found on the headwaters of the Gila by soldiers from Fort West. About $60,000 lies buried at the side of the burned cabin somewhere on the Rio Prieto (Black River). The burned cabin was on the side of a narrow gulch, near an old stone corral. Sluice boxes were in front of the cabin with large flow of water with scattered pines nearby.

(34) "Lost Mine of Don Miguel Peralta" Arizona

About an hour's travel from Phoenix in the foothills of the Superstition Mountains, rich gold ore was milled in arrastres. Senor Jost Ballisteros took charge of the arrastres on Mormon Flat (Now Canyon Lake) for Don Miguel Peralta at Rio Salado. The ore, which came from high in the Superstitions, was in an 18" vein and averaged several thousand dollars to the ton. The source of gold was shut off when Apaches raided the mine, killing all except three Mexicans.

(35) "The Lost Dutchman Mine" 1850 — Arizona

Jacob Wiser and Jacob Walz prospected in the Superstition Mountains. They found placer and searched for its source. Later Wiser and Walz killed two Mexicans who were working the old Spanish Mine which Wiser and Walz had discovered. These men were mistaken for Apaches while near the mine dumps. In 1881, Walz showed up in Florence without Wiser, claiming he was killed while hunting. Wiser, however, was picked up, almost dead, by Pima Indians who took him to the Walker Ranch where he died of pneumonia. Wiser, thinking Walz dead, gave Jack Walker a map of the mine. Walz, in 1882, made a trip to the mine returning with two burro loads—$1600. Walz died on his ranch at Salt River in 1892. Weaver's Needle being a landmark—look for a north and south canyon with a cabin in a cave or shelving cliff.

(36) "Lost Cement Mine" 1841 — Arizona

Three German brothers, all that remained of a party of emigrants that had been killed on the plains, managed to reach California. While resting in a mountain gorge near Mono Lake, one of them saw evidence of work having been done on a vein of cement-like material. Investigation showed it was shot full of lumps of a dull yellow metal, which proved to be gold. Every pound of cement was worth $200. They loaded all they could carry and headed for the Coast, but hard luck again played a part. One brother broke a leg, the second died of thirst, and the survivor was picked up by Whiteman. Whiteman and the survivor formed a partnership and went back to look for the ledge. About the time they located it, they were attacked by Indians and the German was killed. Whiteman, with others, went back later but could not locate the gold.

(37) "Lost Shoemaker Placer" 1877 — Arizona

Charles A. Roderic, old time prospector and former shoemaker, prospected in the Four Peaks Country. There was gold in quantity in ledges and placer. One shovelful produced $23.50. Tonto Apaches caused Roderic and two men to flee for their lives. If you find a shovel and a pick sticking in a pile of dirt, or an old hewn sluice box beside a stream lined with cottonwoods in the Four Peaks country, it is sure to be the "LOST SHOEMAKER."

(38) "Montezuma's Treasure" Arizona

To the south of Ajo Mountain, gold running into the millions was taken from placer mines in Mexico. Many tons of pure gold were picked up from the surface of the ground and in creek beds. This is hidden in the secret recesses of the barren Ajo mountains below Montezuma's Head, near the mining town of Ajo, Pima County, Arizona.

(39) "Shepherd's Lost Mine" 1905 — Arizona

DeEstine Shepherd claims there is $5,000,000 cache of gold here and the mine is still full of gold ore. Shepherd prospected around Tucson for 30 years paying for supplies in gold dust. He died in 1905, at Quincy, Illinois. Some of the quartz showed 75% gold. Three suitcases full netted $12,000 cash.

Go to Tucson, thence 55 miles south, thence west two miles in a gulch draining south.

(40) "Penhachape Mine" 1869 — Arizona

Two Frenchmen left $8,000 on deposit with W. B. Hooper & Co., in Yuma. They are now out of business. It was never claimed. The Frenchmen headed for Eagle Trail Mountain on leaving Yuma. Penhachape Pass produced some ore stacked up beside a trail. The Maricopas are believed to have done away with the French miners near the Bonanza Mine in the Horqua a Hala District.

(41) "Lost Blue Bucket Placer" 1845 — Oregon

An emigrant train upon reaching the gravelly Ford Crossing on the Humboldt River split into two parties; one taking the Humboldt, the other due north to the Black Rock Mountains. Leaving Black Rock they came to a high mountain range. Reaching the top, bearings were taken on the Twin Sister Peaks. In a Canyon on the west slope camp was made at a spring. Chunks of yellow rocks were picked up and several buckets were filled with them by the children. One of the women died here and a blue bucket was hung over the grave on a bush. While crossing the Deschutes River, the wagon capsized, resulting in the loss of the gold except for a few nuggets. The nuggets were later shown at Sutters Fort, California. A party was organized to find the gold. They were ambushed by Indians, only two men returning to California. At Yreka, they told a Dr. Drane of their experience and how to find the canyon.

A trapper later talked to Dr. Drane claiming he knew where the gold could be found. From the head of Goose Lake Valley and at the top of Warner Hill the trapper showed the Doctor two mountains 120 miles to the northeast. The mountain to the right is the one. The Doctor and his companions made a search but couldn't locate the gold. About 20 years later, in 1879, two men traveling across Oregon camped at the agency on the Malheuer Indian Reservation. Adams, one of the men, liked the looks of

the country and decided to prospect, but white men were not allowed to prospect for minerals. The Indian Agent told Johnston & Adams that he had found piles of old rotted timbers, a grave by a spring and a wide deep track down the mountain about three miles from the agency. Mr. Johnson heard the story of the Blue Bucket Placer 50 years after the Agent's story. The Malheuer Reservation was located at the corner of Harvey, Grant and Malheuer Counties and the Agency located on the southwest slope of the Burnt R. mountain west of Beulah and north of Drewsy.

(42) "Lost Mine of the Little Brown Man" — Nevada

Job Taylor, an Indian Trader, gets the little Brown Man to tell him where he gets his nuggets, after practically giving him everything in his store. They went to Susanville, thence on the Honey Lake Road to Deep Hole, Nevada, at a spring. The Indian told him that two sleeps away "catchum gold." In the morning he left to carry a message to the Chief of the Piutes, Chief Winnemuca, and promised to come back. This he did but refused to go farther to show the gold. Later on an Indian boy brought gold nuggets to Taylor who had returned to Indian Valley. Perhaps some Indian will tell an honest prospector where the gold is. It is near Eblings in Virgin Valley, Humboldt County, in a canyon 1½ miles east.

(43) "Geronimo's Mine" — Arizona

While a prisoner at Fort Sill, Oklahoma, Geronimo offered, as a bribe for freedom, to tell of his rich mine in the Verde River Country, in Arizona. The story came from Will James, interpreter at Fort Sill, who tells of a gold mine only one mile northeast of The Bead Mine in a box canyon with three entrances only and the need of rope ladders. On the floor of the canyon is a spring near an old stone or adobe house. This is near Winkelman, Arizona. Indians cashed in nuggets at Tempe.

(45) "Silver Mountain" 1890 — Arizona

As a result of a legend dealing with rich silver veins, Pedro Encinas heads a party of Mexicans in the search, who stop at a San Carlos Indian agency. The veins are more than half silver and little black nuggets of native silver cover the ground. L. K. Thompson of Salt River, a brother-in-law of Encinas, said the mountain was located but was within the Apache Indian Reservation, and although the veins were found, the party returned to Sonora without telling any one of their location.

(46) "The Lost Door Mine" New Mexico

The story is authenticated by records of the Arizona Historical Society and its location is near the corner of Utah, Arizona, New Mexico, Colorado—twelve suns journey from Gila Bend. For three horses and a gun, a Pima Indian promises to lead a party of white men to plenty of gold nuggets as large as wild turkey eggs. If he falls down on his proposal, he is to be shot. Jim Adams, age 34, a freighter; Jay Davidson, John Wingate, Roy Peters and the Dutchman, were among those identified with the expedition. One of these men was panning gold when Adams came into their camp, having just recovered some of his horses that had been driven off by Apaches. The Pima Indian seeing this gold, told of the place where there would be wagon loads.

They crossed the Continental Divide and came within sight of the objective mountains. Crossing a road, the Indian said it lead to a Fort (possibly old Fort Wingate) near the present site of Grants, McKinley County, New Mexico. The trail here crossed a stream lined with cottonwoods.

The Indian led them to a narrow passage in a rock wall below which was a box canyon. A zigzag trail led along the base of perpendicular walls to a clear stream below, which ran through willows and a rocky bed. Excitement ran high when members of the party looked into the stream and saw that the swale contained plenty of gold. Some panned along the banks all night by firelight. The Indian left. Next morning about forty warriors with their Indian Chief were sighted coming down the trail single file. They proved friendly but the party was told by the Chief to stay in the canyon and not go above where the tribe was camped. A cabin was erected and it was agreed that all the gold would be placed in a vault under the fireplace. Deer was plentiful so the Camp had plenty of meat, but provisions were running low. A party, headed by John Wingate, set out for Fort Wingate for supplies. The Dutchman, having accumulated $10,000 in gold, decided that that was enough for him. The arrogant and quarrelsome fellow wouldn't join the others in pooling the gold. More than

a week passed without the party returning with supplies. Adams and Davidson went up the trail and through the entrance to the canyon to look for them. They discovered that the whole party had been killed and the supplies taken by the Indians. Six men were in the party but Wingate's body couldn't be found. About this time the men panning above the canyon were attacked by the Indians and a band of about twenty went down into the canyon and killed the men there and burned the cabin. Adams, Davidson and Wingate (Brewer) were the only ones who escaped. They hid out and watched the Indians spread out over the region in a search for them. Later, Adams and Davidson went back to try to recover the gold but without success. They wandered for thirteen days before being picked up by soldiers from Fort Apache. Davidson died at Fort Whipple. After several days of wandering, Wingate was found by friendly Indians. An old squaw cared for him for about a week and he was then taken to a trail which they said led to the Rio Grande. A pack train took him to Santa Fe. Dr. Spurgeon, having seen Adam's nugget, made a search as did Wingate and others, but none of them found the entrance to the canyon.

(47) "Lost Bandit Mine" 1820 — California

Near Vallecito on the old Butterfield Stage Line, the bandit husband of a Mexican woman buried his loot in several ollas. It consisted of gold bullion, nuggets, etc. His wife tried to find the cache after he was killed, but could never do so.

(48) "Waggoner's Lost Ledge" Arizona

Fred Mullins was driving the Pinal-Mescal Stage with a passenger named Wagoner, who made many trips with him. In 1894, Wagoner decided to make a short cut from a point near Morman Flat to Pinal, his home. From Morman Flat, his route was as follows: Apache Road to Tortilla Flat; southeast up Tortilla Creek for several hours to fairly level country on the east slope of Tortilla Mountain; due south through lower hills to LaBarge Canyon, near its head, the best part of a day's travel from Morman Flat. He missed a trail known to him, being about a mile west of it when he decided to stop for the night. This was in La-Barge Canyon at a spring about three miles due east of Weaver's Needle and about due north of Miner's Needle. Next morning after traveling about an hour, he found a rose quartz vein outcropping on the east slope of Miner's Needle. The rock was literally studded with bright yellow gold. Wagoner's trail from the vein left La Barge Canyon to skirt Picacho Butte on the east, down Red Tank Canyon to Frazier Canyon; thence down Randolph and Whitlow Canyons to the Whitlow Ranch, a point on the stage road. Wagoner made many trips to his find, carrying the rich ore out in a suitcase. He gave the stage driver, Fred Mullins, a map to show him how to find the vein, after saying he was going to take life easy and vanish. Mullins couldn't find the place.

(49) "Lost Arch Mine"
1883 — California

A man named Amsden and a prospector of Needles, set out on a secret mission from Needles to the Turtle Mountains, about 40 miles southeast. "Gold nuggets are to be had for the trouble of picking them up." In adding to the hoard of gold, the men failed to note their diminishing supplies of food and water until they were nearly gone. They took what gold they could carry and buried the rest, which if found, would net a tub half-full of gold nuggets.

Amsden reached Goffs, but his companion didn't make it. Amsden departed for his home in the East, taking his secret with him, but wrote Dick Colton, who aided him at Goff's, sending him a map describing the location. The location was not far from a natural arch. Colton, Mort Immel of Barstow, and Herb Witmire started out to find the gold. No trace of an arch was found by them. Walter Ford of El Centro and John Hilton of South Indio, made a trip to the arch as Hilton had discovered it previously not knowing of the gold story. They found carnelian, plumegate, opal and geodes, but no gold. Their route was—14 miles east from Rice; thence up a draw for 15 miles where the car was parked at a point ¾ of a mile southeast from the arch which is backed by a Butte to the West. Chester Pinkham of Eagle Rock said placer ground exists a few miles north of the arch but not at the arch.

Desert Magazine of January, 1945, shows photo of a large arch, in the Turtle Mountains at Mesquite Spring, near Carson's Well.

(50) "Lost Mines of the Peraltas"
1846 — Arizona

Eight fabulous bonanzas were lost by the ill-fated Peralta expedition from Mexico. Miguel Peralta—owner of the Peralta Silver Mines of Chihuahua, had three sons who were miners, and followed the advice of the elder Peralta in prospecting and mining. The three boys were Pedro, Ramon and Manuel. They prospected the Salt River in Arizona, then came to a virtual paradise, for that country, a verdant valley in the middle of which La Barge Creek tumbled in cascades from a jagged mountain to the south—the Superstitions. On what is now Mormon Flat (Canyon Lake) they successfully panned gold. Later, in order to work gold ore discovered by Pedro, they built two arrastras. Pedro's route to the gold was up La Barge Creek, into Boulder Creek, on up Needles Canyon to what is now Bluff Springs Canyon. On the eastern slope of a black-topped mountain, a mile and a half north of La Combrera, he found rich twin outcrops of reddish gold-bearing quartz. Circling the mountain, he found two more bonanzas. Pedro went back to Chihuahua and in 1848 returned with 68 men and 200 mules to mine the gold. They mined and prospected, discovering more veins. Apaches were seen in the vicinity of the arrastras several times, then after gathering a large force so as to surround the miners, they attacked. The men at the arrastras were killed first, except one man who fled to notify Pedro. Many of Pedro's men were killed before he started his retreat to the west. He carried what gold he could and later buried it as his miners fought a rear guard action. The route was blocked by the Apaches who forced the men back to high cliffs and then killed them all. In 1901, a cowboy found $35,000 amid a heap of Spanish-shod mule bones on top of a black-topped mountain.

C. H. Silverlock and a partner found $18,000 of gold concentrates in the same area. The veins and diggings were covered up by the Apaches and never found.

Manuel and Ramon, satisfied with the results of their placer operations, returned to their home in Mexico.

(51) "Pedro Peraltas Mine"
1860 — Arizona

Miguel Peralta, the father, died at this time. He had continued his mining operations at Chihuahua City and Monterey, and also invested the money Pedro had given him years earlier. Ramon decided to leave his home near Cananea. Manuel, being married, was to remain on his ranch.

Ramon, being single and foot-loose, often visited Cananea—then an active, wild mining camp. Here he met two adventurers, Jacobs and Ludi. Before leaving for Mexico, Ramon gave them a map of the Pedro Mine. After about eleven years the Jacobs-Ludi mine played out, so they decided to look for the Superstition Mine. The map's description follows: Approach the region from the south, enter the first deep canyon from the western end of the range, climb northward over the backbone of the range itself; approach a sombrero-shaped peak; travel downward past the base of the peak into (East Boulder) Canyon, running northward until coming to a tributary canyon on the east side of (East Boulder) it being very deep, pot-holed and densely wooded with scruboak; proceed up the tributary southward (Needle Canyon) until a point is reached where the outlines of the hat-shaped peak to the south and the black-topped mountain to the west both match from the same place the outlines from the map. Near this spot they would find a marker upon the end of a rocky ridge which would be pointing to the mines.

The Apache women had covered up all of Pedro's bonanzas except a shallow one, which was found by Jacobs and Ludi. Here they saw the beautiful rose quartz with about a third shining yellow metal. But others ahead of them had found this rich deposit. Two prospectors were working at the western edge of the Superstitions at what was to become the boom camp of Goldfield. Jacob Walz and Jacob Wiser, while tracing placer gold in Needle Canyon, were camped at Bluff Springs Mountain near the northwest corner. Here they heard miners at work, and upon investigating, found Jacobs and Ludi at their mine. Jacobs and Ludi were shot by Walz and Wiser; then Walz shot Wiser to claim the gold for his very own.

(52) "Lost Dutch Oven" 1894 — California

Thomas Scofield of Danby, California, a prospector of fifty years experience, was working for the Santa Fe Railway on a water tunnel. One Sunday he strolled from camp and came upon a spring at the base of the Clipper Mountain. From here a trail led up a wide draw which led him, after a three hour walk, to an abandoned camp with old blankets about and evidence of tent stakes, utensils, buckets and windlass, near an 80 foot shaft. From the dump near the shaft Scofield took some quartz samples that assayed $5000 to the ton. Darkness forced him to remain until morning. It was then he discovered the old 200# dutch oven containing free gold, nuggets, thin sheets of gold and tin can full of gold dust and some rich quartz. Returning some months later, he couldn't find the spring or the mine. Cloudbursts and similarity of the dry washes could account for the difficulty in finding a claim. During the Spring of 1944, a Mr. Smith of Los Angeles took a mining engineer to the Clipper Mountains to see and pass on a copper claim that Smith had staked out about ten years previous. After a search of three days, most of it, however, put in negotiating the tough roads, they were forced to return without

finding the large copper ledge discovered by Smith.

(53) "Lost Indian Mine of Butte County, California"

About 1884, an old Indian started bringing his poke of gold to town and continued at intervals of four to six months. He was followed, but none of the prospectors could keep him in sight. While on a spree in town, the Indian died. Robert G. Ferguson tells of how F. W. Harrington, with two partners, scientifically combed the whole region, and after two years found the mine. It was a dandy and is producing to date. A slide had covered the Indian's tools and sluicebox, which was to be uncovered about fifty years later.

(54) "Blue Bucket Placer" 1845 — Oregon

A wagon train camped overnight in Eastern Oregon. Children found some bright yellow stones in the Creek. They filled some of the blue buckets hanging beneath the wagon. One of the pioneering women died this day and her

James Marshall first found gold here at **Sutters Mill** in 1848.
— California Department of Natural Resources

grave was marked with one of the blue buckets. Two days travel from here while crossing the Deschutes River, at Shearer's Bridge, (the old toll bridge on the Oregon Trail) most of the buckets were lost. The kids still had a few of the bright stones and it was when the party saw what nuggets looked like at Sutter's Mill on the American River that they realized the kids had found pure gold nuggets. Prospectors have searched for years. The John Day River is the one that could have produced gold. The writer knew an old prospector who went into the headwaters of the John Day by hiking from the mouth at the Columbia River. He would come out every fall with gold amalgam and highly colored stones which he had polished. This was about in the year 1910.

(55) "Ed Schieffelin's Tombstone Ledge" and "Coffee Creek Mine"

In 1877 Ed was on patrol for Al Siebers U. S. Army troop after old foxy Geronimo, leader of the Apaches. While along the San Pedro River he filled his saddlebags with what looked like good silver ore. He withdrew his few dollars, bought some bacon, flour and beans, a second-hand pick and shovel, and went off to the San Pedro Hills. His operating base was the old Frederic Brunckow Mine, first opened in 1858. Brunckow was found dead in 1860 with an arrow in his body. Fort Huachuca was the base for the U. S. Patrol. About two miles from his camp, the draw Ed was in forked. A small stream of alkali water and a jack rabbit led him up the draw to the right. Late that day he dug his pick into a likely piece of ground. When he withdrew the pick, the sun shone on what looked like pure silver and was soft enough to take the print of a 25 cent piece

that he carried. Al Sieber had told Ed that all he would find was his tombstone, so Ed called his find the "Tombstone." With his brother and a mining engineer at the McCracken Mine, a Richard Gird, three claims were taken up. The ore samples assayed from $40 to $2000 a ton. This was the "Tough Nut Lode," a $75,-000,000 producer. Ed and his brother received one million each. Gird promoted and developed the property, taking his reward in millions.

In 1897, Ed left for Douglas County, Oregon, to look over Coffee Creek where he panned for gold as a kid. That same year in May, Ed was found dead in his cabin by a deer hunter. It

was probably heart trouble. He was 49 years and 8 months old. With his Will, Ed left his nephew a map and a problem to solve in finding the gold mine he discovered, which was to make the "Tombstone" look like nothing in comparison. The nephew was killed at Verdun and a pal received the map. Many have searched for the mine but only Ed's ghost knows where it is.

(56) "The Silver King Mine" 1870 — Arizona

Due southeast of Arizona's Superstition Range near General Stoneman's Camp at Picket Post, a soldier named Sullivan was on an Apache's trail. Resting, he picked up a handful of stones, among them black silver ore of bonanza richness. Sullivan showed his find to Charles G. Mason, a rancher on the Gila River. On March 22, 1875, Mason staked the fabulous Silver King Mine. Sullivan had refused to divulge the location.

Black sulphide ore ran $20,000 a ton. Mason and partners shared a million dollars in two years. Picket Post became Pinal. The mine continued to produce for several years.

(57) "The Lost Soldiers Mine" 1875 — Arizona

Two soldiers, following their enlistment, took the military trail through the Superstitions. Traveling southeast from Mormon Flat and crossing Salt River, they entered the mountains between Kayhatin and La Barge Creeks, past Indian ruins in Garden Valley, over Black Mesa to West Boulder Canyon at this point, up East Boulder toward Weaver's Needle. Then east on the trail to pass below southern slopes of the black-topped mountain which separates East Boulder and Needle Canyons, a half mile apart. Flushing a deer, they fired on it and came upon a reddish vein about a foot thick halfway up a black-topped hill. When shown the samples, Jack Frazer, Foreman of the Silver King Mine, called in Mason the Superintendent. The few samples were weighed and netted the former soldiers $500 in gold. Mason staked the men who set out to backtrack themselves—followed by a peg-legged gambler on a horse. The men were never seen alive again. The gambler Smith disappeared, later showing up in Alaska with his bags stuffed with glittering gold. The pasture of the Quarter-Circle-V Ranch showed the remains of the two soldiers who were ambushed. Smith tried to get Arizona friends to stake the mine from his description, but it was never found.

(58) "The San Saba" 1830 — Texas

Spanish gold and silver by the millions is stored in caves near Calf Creek in McCulloch County, 25 miles East of San Saba Fort or on Jackson's Creek, six miles east, which is a tributary of the San Sabo. Here a fight took place and the Spaniards, after hiding their wealth, were killed or driven from the country.

Wes Burton claims the Indians didn't show Bowie a mine but 500 jackloads of pure silver stored in a cave by the Spaniards. The location is supposed to be three leagues (11 miles) up the San Sabo River to the west from the old fort, then up Silver Creek one league. The old fort is near Menard. Recent figures credit the 500 jack loads of silver as worth $13,000,000 and add $1,500,000 of gold. A search is being made near Boldthwaite, Texas, for the Bowie and San Saba mines.

(59) "Gold Mines of the Nueces" 1762 — Texas

Mission San Lorenzo de la Santa Cruz on the Nueces near present Camp Wood. Map shows a Silver Mine on the San Saba and one on the divide between the Nueces and the Frio. General John R. Baylor commenced his search that cost him a fortune just after the Civil War. In 1906, Henry Yelvington made a search, but finding deer, turkeys, javelinas, fish, bee trees, foxes, bobcats, squirrels and bear, did not mind missing the gold. In 1915, float assaying $116 in silver and lead near the ruins at Camp Wood, was found by George Baylor, a son of the general. The old tunnels show only low grade ore but there is more exploring to do.

(60) "The Rock Pens" — 1873

Six or seven miles below the Laredo Crossing on the south side of the Nueces River near the hills, there is a tree in the prairie. Due west from that tree at the foot of the hills at the mouth of a ravine, there is a large rock; under the rock there is a small spring; due east from that rock is a rock pen and due east a few yards there is another pen of rock. In that pen is the spoils of thirty-one muleloads. The Laredo Crossing is not at Fort Elwell but on the Henry Shiner Ranch in McMullen County. The rocks have very likely been scattered. This treasure, supposed to be the spoils of bandits, has never been found. The name Guidan Pasture, which joins the Shiner ranch, is mentioned as the location.

(61) "Sin Caja (Without Coffin)" 1875 — Texas

John Fogg in Corpus Christi was told by a Mexican that he had been held in prison for 37 years after having been accused of desertion from the Army. As a cargo of silver was going to Mexico City from the Saba Mines, when a boy he had enlisted as a guard. They were attacked by Indians. Caching the silver between some rocks, a defense was set up, the Mexican boy going down the ravine to look for water. All of the party was annihilated except him. Guided by the Mexican boy, Fogg and two others, traveled three days from Corpus Christi, and camped that night in sight of the mountain. Eating fresh venison that night for supper, the guide was seized with cramps and died without help in locating the treasure; the others gave up the quest.

(62) "Fort Ramirez of the Rancho del Ajo de Agua Ramirena" of the Nueces Country — Texas

About 1813 Indians preyed on the country causing desertion of old ranchos on the Ramirena Creek. Tol McNull dug up $40,000 of the money buried there.

(63) "Karl Steinheimer's Millions" Texas — 1827

Having mined in Mexico for ten years, Steinheimer joined in a revolt against Mexico. Traveling the San Antonio Road he threw in with a Mexican, Manuel Flores and his soldiers, as protection against the Apaches. The soldiers being attacked by Texans, Steinheimer separated himself from the military party and set out for the Colorado River. In the hill country, sixty or seventy miles north of Austin at a place where three streams combine into one, he unpacked his ten burros and concealed the freight. A large brass spike driven into an oak some forty feet away designates the spot. This must be at the Salado, the Lampasas and the Leon near Belton, which forms the Little River. About twelve or fifteen miles southeast of the junction a prominent bunch of "knobs" overlooks the tortuous valley of Little River. For 75 years the quest has been prosecuted. Steinheimer was wounded at this place and died later, after leaving a letter and description for a lady in Saint Louis.

(64) "Midas on a Goatskin" — Texas

The old Spanish Trail crossed over into Texas from Mexico at the mouth of the Pecos River, then east, circling Seminole Hill just west of Devil's River, across Mud Creek and thence to San Antonio and New Orleans. It was the route used by the Antiguas for carrying gold and silver out of Mexico.

The country was still full of dead Spaniards and of bullion and bags of money. Seminole Hill hides a lot of treasure. In 1880 Dee Davis of Sabinal tells of his Uncle Ben and father making the crossing at Mud Creek on the old Spanish Trail. They discovered three metal bars exposed after a severe rain. A visiting surveyor, finding the bars under the bed, asked if he could take them to town for assay. His delayed report showed babbit, but the surveyor soon acquired a fine home and much land. Dee Davis made one try for more bars at the point several years later. He said when his boy got old enough to shift for himself he will make several investigations. Among others, he would like to look for that cave in Seminole Hill where an old Mexican pastor named Santiago stepped right on top of more money than he'd ever seen all put together, some stacked up and some just lying on the floor of the cave. Square Spanish coins. But what Dee Davis would rather get at than Santiago's cave is an old smelter across the Rio Grande in Mexico. just below the mouth of the Pecos. It's under a bluff fronting the river. The mine was below the mouth of the Pecos. It is covered by a bed of gravel. A scraper will uncover the mouth of a shaft. Then this "El Lipano." The Lipan Indians worked it. They pounded out gold ornaments in rock tinajas across the Rio Grande from Reagon Canyon. Devil's River and Painted Canyon are the forks where Indians hid maletas of money between the forks and the long bluff on the south side of the Rio Grande below the mouth of the Pecos. Also the "Lost Nigger Mine." Go to that round mountain down in the vegas on the Mexican side just opposite the old Reagan Camp, called El Diablo, also Nigger-head or El Capitan. Halfway up on a two-acre mesa against the mountain wall is a chapote bush. Under that is a hole with an old ladder with rungs tied on with rawhide. Hidden behind the chapote is a macapal (ore basket).

(65) "Lost Nigger Mine" — 1884

Look in the Big Bend country, either on the Texas side above the mouth of Reagan Canyon in Brewster County, or in the Ladrones (Robber) Mountains of Mexico. From the Southern Pacific Railroad it is 75 miles with only one house between. The four Reagan brothers moved their cattle from the Pecos to the Rio Grande. The country was all open. Headquarters became Reagan Canyon Camp. A Seminole negro, Bill Kelley, was hired at Dryden. In 1887, Seminole Bill announced while the outfit was sitting around a campfire, that he had found a gold mine. Bill offered to show Reagan the mine which was only about a half mile from the meeting place of the two after spending most of the day from camp looking for lost horses. Reagan refused to go with Bill, telling him, "we're not feeding you to hunt mines." Bill sent a sample of the ore to Lock Campbell, a conductor on the passenger train between San Antonio and Sanderson. Bill took a horse and left the Reagans at night, reaching Stillwell Ranch at the Huerfanito in northern Coahiula, a day's ride south of Shafter's Crossing. He showed Stillwell, a former engineer, some of the ore, who recognized it as very rich gold ore. Campbell sent out a prospector to look for the mine, but this man, being afraid of bandits, decided to prospect in the opposite direction, going up the river to Terlinqua. He found rich cinnabar. Campbell's partner told the prospector to let it alone as cinnabar was no good. The owners of the quicksilver mines at Terlinqua became rich. Seminole Bill's ore assayed $80,000 a ton, according to Campbell when interviewed in 1926. The Mine is on the Mexican side of the Rio Grande in a canyon among rocks sticking up so close to each other that a horse couldn't get to it. The Reagans searched for the mine but couldn't find it. Seminole Bill was never found either. There are granite outcrops in the Reagan Canyon Country, and the Chisos Mountains are well mineralized. John Young of Alpine and Jack Haggard searched for the mine. John Finky, a prospector for Campbell and others, found the mine but never got back to it.

(66) "The Engineer's Ledge"

Hughes, the driver of construction engine on the Southern Pacific liked to pick up rocks. A black ledge over a ravine not far west of Paisano Pass, just south of the railroad, a cluster of sotol, a gnarled pinon in the ravine, a spring ten miles up the ravine would give piped water. Hughes, while in Denver, had some of his rocks assayed. One was very rich and he thinks it came from the black ledge. Returning to the place, he couldn't find any similar rock. He decided he didn't know just where the rock came from but searched over the old construction territory for years without locating his ledge.

(67) "The Lost Padre Mine" 1659 — Mexico

An Indian goes out from the Church of Nuestra Senora de la Guadalupe in Juarez, across the river from El Paso. He returns regularly with a small cargo of gold and silver ore. The entrance to this mine is supposed to be in Franklin Mountain, which can be seen from the Church across the Rio Grande in Texas. In 1888 an old government packer named Big Mick and a man named Robinson located the Padre Shaft. They were backed by a man in Santa Fe who believed the Jesuits had secreted three hundred jack loads of silver bullion at the bottom of the shaft. They found no treasure.

(68) "The Breyfogle Mine"
1862 — California

The report of a rich silver strike at Austin, Nevada, caused three men in Los Angeles to start out afoot for the Reese River District. Two of the men, McLeod and O'Bannon, were killed by the Indians. Breyfogle escaped, travelled from the point of ambush, a spring in the Panamint Mountains, on the eastern slope at a rock tinaya on a crude Indian trail to the floor of the valley, then crossed Death Valley the afternoon following the Indian attack. He found an alkali spring, the water making him very sick. An hour into the foothills from this spring, he stayed the night, being on the west slope of the Funeral Range. A green spot about halfway up the mountain and to the south attracted his attention. It was about three miles away. About half this distance he found a soft greyish-white rock showing free gold. Then the vein of a pinkish feldspar rock much richer than the float was found.

The green spot proved to be a mesquite bush, the uncooked beans making Breyfogle sick. From here he covered 250 miles, crossing the Amargosa Desert to Baxter Springs. In Smokey Valley he came across rancher Wilson who took the big six foot, almost naked Bavarian to his house. He recovered his normal functioning of mind and body and went to work for Jake Gooding at Austin. Gooding, George Hearst, and Donald F. MacCarthy all searched, aided by Breyfogle, but he couldn't guide them to the vein. Months having elapsed since he returned to the spot, it is supposed a cloudburst covered it up.

(69) "Yuma's Gold" — Arizona

A graduate of West Point, a young lieutenant, was acting quartermaster of the Post at Fort Yuma. Dishonest dealings caused him to take refuge with the Yuma Indians. The lieutenant was called Yuma. The Post is about 60 miles northwest of Tucson, Arizona; where the Arivaipa Creek joins the sand bed of San Pedro River, is old Camp Grant. About 10 miles distant in the Arivaipa Hills, the Arivaipa band of Apaches had their main camp. The chief, in consideration of a rifle, beaded belt with ammunition and silver spangles,

promised to show Yuma the source of the Apache gold. Travelling in a northerly direction from the Arivaipa Camp, they ascended a long ridge on which they kept on going about three miles, coming to a crest overlooking San Pedro Valley to the east. About six miles more to the north, while along a gulch, the chief stopped and pointed to a six-foot crater-like depression. Using a hunting knife, Yuma broke off a handful of rose quartz wonderfully rich in gold. The gulch headed a few hundred yards beyond in very rough terrain. Yuma enlisted the aid of a kin of John J. Crittenden. They went into the mine area from Tucson; leaving late one afternoon, they reached Camp Grant the next morning. The next night they travelled northward down the San Pedro some ten miles to a point opposite the mine. At daybreak they climbed the range to the west and after a few hours came to the gulch. Taking about thirty pounds of the ore, they coursed down the western slope of the mountains and then crossed a trackless basin. Travelling all night, they reached Tucson about daylight. They recovered about $1,200 from the thirty pounds.

Yuma and his wife were killed by the Papagos; Crittenden by the Arivaipa Apaches, and they in turn were massacred in 1871 to every last man, woman and child, in what is known as the Camp Grant Massacre. The desert guards the rose quartz.

(70) "In the Sunshine of the Pecos"
New Mexico

Pecos Village, thirty miles southeast of Santa Fe, New Mexico, lies seven thousand feet up. When Coronado traveled the Santa Fe Trail from Missouri it twisted by Pecos Village, then known as Cicuye. A church was erected in 1620. Jose Vaca lived in the modern Pecos Village down under the hill from the ancient pueblo. Jose tells of the Mines:

(71) "La Mina Perdida"
New Mexico

"We hunted deer far up towards the head of the canyon. We had a deer loaded on a horse when it began to rain. Taking shelter under a tree we found an old flume and an anvil. Nearby is a canyon with pretty water for fish. Next time for hunting we go back to look for flume and anvil but could not find

it. Surely it was the Lost Mine watched over by the Indians."

"Also on the headwaters of the Pecos we came across a whole mountainside of mica. A partnership was formed with Road Construction Boss, Mr. McCarch. In July, when it was not too cold, a man named Iglehart is sent in McCarch's place. Three days' travel from the Pecos River takes us into those high mountains." A bad heart caused Iglehart to sleep all the time. The mica claim was not located.

(72) "The Montezuma of the Pecos" New Mexico

It is the custom of the Pueblo Indians to strip a man naked, stake him down in an ant hill and leave him there until the ants had left nothing but bleached bones, in case he has been accused of rape. A young brave was jailed, but freed by the Sheriff Don Solomon, when the brave told his story. In gratitude, the sheriff was told where to find the gold. Go to the old church at Pecos pueblo and find the Spanish road to Santa Fe. On the hill near the pueblo is a white rock with an old cross. Next to that is a black rock which hides a cave. Dig under the white rock. Seven feet under the patron (the dead man who guards the gold) is the gold itself.

(73) "Jose Vaca's Cave" New Mexico

His cave is on one of the Tecolate Mountains overlooking the Pecos River just below the village of Pecos. A derrotero, brought from Las Vegas, has been the basis for most of the work done in the cave. This derrotero called for the rocks and crosses before the cave. The legend says the early Spanish were bringing an immense cargo of pure ore from Mina Perdida up the Pecos. They were attacked by Indians and retreated to the cave, stored the ore, and sealed the cave.

(74) "The Crossings of the Pecos" Texas

Far south of the Santa Fe Trail three other transcontinental routes crossed the Pecos converging west of it: the Butterfield, or Southern Overland and Emigrant Trail — Saint Louis to California; the Chihuahua Trail from New Orleans to the west, and between these two the San Antonio - El Paso stage road. The Horsehead Crossing was the most famous of the three, the other two being Fort Lancaster and the Pontoon. Bones of cattle used to line the waterless Good-night Living Trail stretching for ninety-six miles from Horsehead Crossing to the head of the Concho River. Every spring and watering place has its history or anecdote connected with Indian violence and bloodshed starting with the diary of J. M. Bell, 1854, to Edward F. Beale in 1857. Great trains of wagons passing between Chihuahua City and San Antonio were veritable argosies of treasure. In 1876, August Santlehan transported 350,000 pesos in Mexican silver. Theophilus Noel records that "the ore from all the mines of North Mexico was hauled to San Antonio to be taken to Fort Lavaca by Texans on wooden wheeled carts and then to England." Some of it is at Castle Gap.

(75) "Maximillian's Gold" — Texas

A caravan of wagons coming out of Mexico in charge of an Austrian is met by six Missourians, former Confederate soldiers who swore they would not live under Union rule. The Austrian hired them to guard his cargo, which he said was flour. His curiosity aroused later, one of the men investigated and discovered a fortune, fabulous in value, consisting of Spanish, Austrian and American coins, vessels of gold, silver and some bullion. At Castle Gap, fifteen miles east of Horsehead Crossing, the Austrian and his companions were killed. Papers taken from a chest revealed that the leader was a follower of Maximilian's, carrying the royal fortune out of Mexico to Galveston. Noticing the landscape of rock, sands and lake, the robbers buried the treasure and rode on east. One remained behind at Fort Concho because of illness. The others rode on and were attacked by Indians and killed. The sole owner of the fortune decided to enlist the aid of the James boys in disposing of the fortune. Near Denton, he camped one night by chance, with some horse thieves and all were made prisoners by a sheriff's posse the next morning. The Missourian, in a dying condition at the Denton Jail, left a plat to Doctor Black and lawyer named O'Connor. When they reached Castle Gap months later, the landscape had changed and the landmarks could not be identified. They found some wagon irons marked by fire.

(76) "Rattlesnake Cave on the Pecos" Texas

To the Southeast of Castle Gap, the Chihuahua Trail followed up Devil's River, left it at Beaver Lake, followed Dry Draw and crossed the Divide to the Pecos. Somewhere along Dry Draw was Rattlesnake Cave. A sixteen-year-old Mexican pastor, while herding sheep out from Beaver Lake, discovered a cave, the entrance the size of a barrel. He wriggled through only to confront the largest rattler he had ever seen. Other rattlers joined in the buzzing and the Mexican beat a retreat. Stopping to pick up a couple of rocks to throw at the rattler, he heaved one, killing the snake, but found himself carrying the other as he left the cave. The "rock" proved to be a crude block of gold and silver mixed, weighing about 3 pounds and seven ounces ($1100). The boy's mother, who lived in Ozona, and also the padre, advised the boy not to tell the secret. He was told that the snakes were spirits guarding the treasure. Wes Burton followed directions given him by Neal Russell, a Ranger. Neal had died, so Burton and Preacher Crumley headed for Dry Draw, about thirty miles out of Ozona. A gully nearby marked the old Chihuahua Trail. While in this gully, a stranger appeared, who said his name was Cox; he and another man hearing rattlers in a cave nearby had tossed a stick of dynamite into it blowing out a wagonload of them. Burton and Crumley went to town for grub and a new shovel, but sickness and other troubles tied them up for a year. When they got back to Ozona they heard that a man named Cox had bought a 50,000-acre ranch down Howard's Draw, had it well stocked and money in the bank.

(77) "The Secret of the Guadalupes" Texas

The tradition of gold in the Guadalupes runs back to 1680. An Indian of Tabira conducted Captain de Gavilan and thirty other Spaniards to a wonderfully rich gold deposit on the eastern spur of the Guadalupe Mountains — called Sierra de Canizas Mountains. They returned loaded with nuggets and ore. Guadalupe Peak, highest point in Texas, is 9,500 feet above sea level just below the New Mexico line. The Mountain chain extends nearly 100 miles northerly from the Rio Grande. Old "Ben" Sublett, William Colum Sublett, was a famous seeker of gold in the Guadalupes. He lived in a tent near the town of Monachans while the Texas Pacific Railroad was building. Ben would leave this town in West Texas often to prospect; then moved to Odessa. After many trips, he reported the finding of a mine so rich he could build a palace of California marble and buy up the whole State of Texas as a backyard for the children to play. Sublett said he would carry his secret to his grave. However, he did offer to take a friend to the mine, a man named Stewart. This was while Stewart was acting as guide to some railroad people. Stewart couldn't leave his party, but went with Sublett as far as the top of a blue mound to the west where Ben tried to show him the mine through a long spyglass. Upon Ben's return to Stewart's camp, he had a small sack of large nuggets. Sublett died in 1892. Stewart looked for the mine, as also did Ben's son, Ross, who had been to the mine with his father when the boy was nine. Ross thought the mine was within 6 miles of a spring in Rustler Hills which are about forty miles east of Guadalupe Peak.

(78) "Gold Is Where You Find It" (Various Locations)

Also silver, lead, zinc, copper and quicksilver. In 1545, an Indian hunter discovered the famous Potosi silver mine when he clutched a bush to prevent falling and found glittering silver clinging to its roots.

The Comstock lode was located (California) when in 1859 a half-witted prospector "Pancake" Comstock scooped some dirt out of a gopher hole, sprinkled with gold and silver.

Don Scribner on Wood River (Idaho). While climbing a hill to look for lost horses, he saw a badger hole about half-way up. He examined the rocks thrown out by the animal. Soon he uncovered three feet of solid galena running high in silver. This was the Minnie Moore which sold out for half a million dollars.

(78) At Tonopah, Nevada, Jim Butler's mule kicked off a shallow cap of rock that hid one of the richest of recent mines.

In 1927, a badger hole was responsible for Horton's rich pay dirt in the Malapi, north of Tonopah — the Weepah stampede resulted.

A squirrel on the Sabinal River in Texas, burrowed deep enough to expose some quicksilver.

(79) "The Lost Mine of the Caddos" Texas

This is supposed to be the source of the lead that the Caddo Indians took to Shreveport to trade. A foxhunter, while going up in John's Creek (a tributary to Caddo Lake) noticed considerable debris at the mouth of a crude shaft with some pieces of galena.

— 80 —

James Goacher — 1835 — settled on Rabb's Creek near Giddings, in Lee County, (Texas). The road to his place from Austin was known as Goacher's Trace. Settlers bought their lead from him. Indians killed Goacher, his two sons and a man named Crawford. One possible clew to Goacher's lead mine was the finding of a solid piece of lead by a Mrs. Johnson when she and her husband were fishing on Rabb's Creek somewhere between Two-Mile Crossing and Five-Mile Crossing. This chunk of lead was used for years as a door stop before its worth was discovered.

— 81 —

About 1850 an old Dutchman named Frank Vanlitsen, who lived near Wallace Bridge, on the Lavaca River supplied settlers with lead. He never divulged its source. North of Sabinal a rancher named Hoffman also sold lead. He was killed by Indians and never shared his secret of the lead.

— 82 —

Texas — a buyer of horses, Thomas Logest, in 1887, while riding between Salt Fork and the Double Mountain Fork of the Brazos and the trails in the general direction of Croton Creek, found an old pick and shovel under a canyon bluff. This lead to a ledge of rich lead ore. This is in either Stonewall or King County near Kiowa Peak.

(83) "Copper On The Brazos" Texas

Somewhere near the junction of the Salt and Double Mountain forks and within sighting distance of Kiowa Peak, Spanish miners carrying a large quantity of gold from Mexico, looked for the copper mines of the Brazos known by the Indians. They found the copper, but ran into hostile Indians, buried their gold and left the country. A plat was left with a faithful mestizo. This was passed along for three generations and turned up in 1908 in the hands of an American who turned up in the town of Haskell. A large party did extensive digging and searching for most of a year. Many copper and silver ornaments, beads, gold buttons, etc., were found, but no treasure. Two heavy copper vessels were found and the conclusion was reached that a Mexican had taken at least a part of the treasure from the containers.

(84) "Moro's Gold" — Texas

About 1833, some young caballeros in the State of Tamaulpias organized the Moros, (picturesque robes of the Moors were worn). Fannie Ratchford's mother, a descendant of one of the Moro Chief's lived with her father, Preston R. Rose, on a plantation called Buena Vista on the Guadalupe River. A Moro, exhausted after losing his pack mule, asks for hospitality for the night. He remains for a long time distributing gold coins and was very lavish with other gifts. This Moro claimed there was treasure lying within fifty feet of the plantation house. Ten years after the Civil War, many were asking permission to dig for the Moro gold on the Rose Plantation.

(85) "The Mystery of the Palo Duro"

Jesus Ramon Grachias was born in Lighthouse Canyon, a tributary of that strange cleft, across the Panhandle in Texas known as Prairie Dog Creek, Red River or Palo Duro. Jesus' father left Texas with other Mexicans when Texans whipped Santa Ana. He set out for Santa Fe. They camped in Lighthouse Canyon, well protected; made friends with the Comanches and found game everywhere. About 1849, he was married in Santa Fe. They loaded a wagon and returned to Lighthouse Canyon. He died in Lighthouse Canyon in 1854. A few days before he died he talked to his wife about fights and gold coins. His last words were "buried fifteen feet east of an old cedar tree." The wife, being partly deaf supposed he meant that he wanted to be buried east of the old cedar tree. She buried him fifty feet east. Jesus had promised his mother that sometime he would put a cross on his father's grave. This he did. Being too young to remember the place, his mother had described it to him thus: A great pillar projects from the center of the canyon like a lighthouse; there are strange markings on the rocks; there is a fine spring of water with old cedars around, herds of antelope and buffalos winter there, panthers and lobos cry out in the night, the letters on a stone at the grave are J. R. G. With the aid of a Mexican who trapped lobos for the J. A. Ranch, Jesus located the grave. He said while there, a dream told him to dig fifteen feet east of the cedar, which he did. He found a chest of old Spanish coins dated before 1821, the year of Mexico's independence. The bank in Santa Fe gave him $7,600 for them.

(86) "Montezuma's Treasure" Texas

The son of a foreman for the Southern Pacific Railroad who lived in a camp several miles east of Del Rio claimed that Montezuma's treasure was not buried in Sugar Loaf Mound at Del Rio. He said that in the hills near their camp between the railroad and the Rio Grande, he and his brother saw an arrow on a rocky ridge pointing southward. Later another one pointing downhill. Then later a ring of rocks about fifteen steps in diameter. When the boy told an old-timer what he had seen, he was told that treasure was undoubtedly buried in the ring of rocks. Going back with the old-timer, he couldn't find the arrow on the rocks.

(87) "The Hundred Million Dollar Trail" 1860 — Arizona

La Paz, born of gold, nurtured on gold, lived and died on gold. Most of the gentle hills lying back of the town were alive with nuggets. One weighed forty-seven ounces. At its docks landed river steamers that brought in the vast supplies needed by placer miners.

Pauline Weaver, guide, scout and pioneer dealing with Indians at Yuma, heard of gold at La Paz. He and his companions brought back $8,000 in nuggets and dust from only a small operation. About one hundred million dollars in gold were taken from the country supplied from La Paz. The old trail eastward from the Colorado leaves La Paz — now Ehrenberg, (La Paz being washed out by flood waters from the Colorado) — paralleling the present Highway 60; goes between the Harqua Halas and the Harcuvars; thence to Wickenburg, Congress and Octave Mines and the Rich Hill placers. From the river area around La Paz came twenty to forty millions in placer; from the North Star to the east, seven millions; the King of Arizona about twelve millions; the Harqua fifteen millions; the Congress thirty millions; Rich Hill produced untold millions with obscure records.

(88) "The Harqua Hala" Arizona

Has several stories as to its origin. An Indian is supposed to have discovered it. Frank Alkire of Phoenix, when a youth, was riding in search of straying cows. At a crossing of a dry wash, he came upon three white men badly in need of water, who upon receiving it, showed Alkire several large nuggets. Needing equipment to work the claim they had staked, they offered Alkire one-third interest for $300. This offer was refused by Alkire as he wasn't interested in mining. This property produced $15,000,000 and was then sold to an English syndicate for $5,000,000.

(89) "The Vulture" — Arizona

In sight of the old freighting trail near Wickenburg, rises a red granite uplift that is a countryside landmark. At its foot is the Vulture Mine. Its name was applied by Henry Wickenburg. Here he found loose rock that contained gold visible to the naked eye at almost any spot across a 30-foot ledge. Gold miners who rushed to the area carried out ore in in tin cans, some of it being 50% pure gold. Wickenburg built the first arrastra in 1864 for recovery of the gold. Michael Goldwater built a small mill; in its first month of operation, it produced $3,000 in gold per day. The Apaches were numerous in the area and scalped many of the travelers. For safety, Henry Wickenburg drove a 160 foot tunnel from his house into the hillside nearby. Wickenburg sold his claim for $100,000 but became involved in a lawsuit over payments. At the age of 86 he took his own life, dying almost penniless.

(90) "The Congress" — Arizona

Stories of the "Golden Cup" and the "Lost Nigger Mine" are brought to mind when Congress Junction is mentioned. The dump of the old Congress property is visible from Congress Junction on U. S. Highway 89. Its shafts were sunk to a depth of 3,900 feet and produced thirty million dollars.

(91) "The Yarnell Mine" — Arizona

Henry Yarnell received a weekly stipend to keep down the dust in Phoenix by driving an old wagon of very substantial construction, with a large zinc-lined, heavy rough lumber tank, which was provided with a circular opening at the top for the entrance of the canvas covered hose for filling, and at the rear, a system of half circular sprinklers. When Yarnell's business expanded to the point where he could take a vacation, he and his wife hooked up a team of horses to the old family buck-

board and hit out for the cool and fragrant smells of the mountains. They camped on the east side of Weaver Mountain, hobbled the horses and turned to setting up camp and getting the evening meal which held so much promise. The morning search for the horses found one of them missing and the apparent distance it had travelled called for a spy-glass. Climbing to a well-elevated ledge for a view, Yarnell accidentally struck the ledge with the point of his telescope, knocking off a piece of rock. It exposed a large nugget of gold. The property produced $300,000.

(92) "The Octave", Red Cloud", "Castle Dome", "French", "North Star", "Swansea", and the "Socorro" — Arizona

The mines were mostly of the glory hole type, esily mined and producing quickly, because of the richness of the ore. This group produced tens of millions of dollars in gold.

The old prospectors are still ranging the hills around Quartzsite and bringing in little glass tubes full of gold to trade for provisions. Spaninsh soldiers garrisoned the Mission near Yuma and were destroyed in the Indian uprising of 1780. Lead was mined in what is now the Apache mine in the Dos Picachos Mountains. Had the Conquistadores with Coronado in 1540 scouted along the "Hundred Million Dollar Trail" in a northly direction from Nogales instead of to the northeast, they would have found their gold. Their search for the fabled Cities of Cibola took them to the Zuni and Hopi Villages.

"THE MOTHER LODE OF CALIFORNIA" (93) "Mariposa"

Spanish for Butterfly — is situated in the old grant given to General John C. Fremont by Governor Alvarado in 1847. Its files contain a wealth of information on the pick-and-pan days of 1849.

(94) "Mt. Bullion"

Was named after Old Bullion, Senator Thomas H. Benton.

Mt. Bullion, once a thriving community.
— Stockton Chamber of Commerce

(95) "Mt. Alphir"

Only a ruined Trabucco store left.

(96) "Hornitos", (Little Ovens)

The birthplace of Ghiradelli's fortune, the Chocolate King. Lumber for the Hornitos Hotel was brought around the Horn. Hotel Hornitos was built in 1860.

Scenes from Hornitos where Ghiradelli's fortune had its origin.

The Plaza

D. Ghiradelli store

Wells Fargo building

The Post Office (former dance hall)
— California Department of Natural Resources

(97) "Bear Valley" — 1851

Established by General Fremont who had the fabulously rich Mariposa Mines. Senator Benton owned the Bon Ton Bar.

(98) "Benton Mills"

Just out of Hell's Hollow, consisted of a dam, power project and mill. They were built by General Fremont.

(99) "La Grange"

Stamping ground of Bret Harte with Red Mountain of Smith's Pocket as the locale for "M'liss".

(100) "Wheeler Mansion"

Employed by Wheeler in his saloon at Coulterville, a pretty and popular dancer won his heart. Having made his pile from the gold fields and his saloon, he could offer his prospective bride anything she desired. She asked for a mansion, the best in the Mother Lode. Construction was started, but before its completion, the girl jilted Wheeler.

(101) "Coulterville" — 1851

Just a shadow of a once prosperous town with its "Hangman's Tree"

Jail still standing erect in **Coulterville.**
— California Department of Natural Resources

Walls and facade of **Coulterville Hotel** remain upright.
— California Department of Natural Resources

(102) "Groveland", The First "Garrote"

(103) The Second "Garrote"
On Oak Flat Road.

(104) Big Oak Flat
The site of a huge oak measuring eleven feet in diameter, which was uprooted in the search for gold.

(105) "Priest's" — 1865
Used as a base of supplies by prospectors. Gold was traded for food and whiskey. At the foot of Priest's Grade near the Tuolomne River was Stevens Place, stopping point for miners from the Bay Region. From the highway abandoned mines may be seen.

(106) "Chinese Camp" — 1854
Bret Harte's "Salvado". The Celestials worked these diggings very thoroughly, but recently a $200 nugget was found there. Table Mountains was the scene of "Tong War" in 1856. It can be seen along the Sonora Road south of Yosemite Junction near Stone Fort.

Store with iron doors, brick front, stone sides, in the **Chinese Camp.**
— California Department of Natural Resources

(107) "Bryne's Ferry"
At the foot of Table Mountain at Stanislaus River.

(108) "Knight's Ferry" — 1849
Trading post on the Stanislaus River, was founded by Wm. Knight, a physician, scout and fur trader.

(109) "Sandy Bar"

(110) "Red Gulch"
Scene of Bret Harte's "Two Men from Sandy Bar and Red Gulch."

(111) "Tuttletown" — 1848

(112) "Rawhide"
Where lived Mark Twain's "Chaparral Quail."

113) "Jamestown" — 1848
Founded by Colonel James of San Francisco.

(114) "Quartz" and
(114a) "Stent"
Both rich mining districts. At Quartz lived Mrs. App, one of the six Donner girls who survived the Donner tragedy.

(115) "Volpone Ruins"
Here was buried gold of considerable amount which was taken from the claims of two Frenchmen. They buried their gold before leaving for the strike at Carson City. Upon their return they found the vegetable garden of the Volpone Brothers over their hiding place, but no gold.

(116) "Sonora" — 1848 "Queen City"
Settled by Mexicans. The Bonanza Mine was operating in 1870 as a very rich producer, turning out $300,000 in one week of operation. Wood's Creek district still produces nuggets.

(117) "Shaw's Flat in 1849"
Jim Fair of the Fairmont Hotel operated here as did John B. Stetson. Fair and Mackay were miners at this time. Later they bought shares in a very profitable mine in the Mother Lode which made them millions.

(118) "Squablestown"
A place of frequent bloodshed.

(119) "Sawmill Flat"
Joaquin Murieta (Robin Hood) dealt "Monte" here in 1852.

(120) "Yankee Hill"
A nugget weighing 249 ounces was found here, valued at $3600. Scene of many diggings.

(121) "Columbia" (American Camp 1850)
Boasted 143 Fargo banks with a capital of $1,500,000, and a population of as high as 40,000. Finding a large nugget was responsible for its growth. Even today plowing in the fields exposes nuggets. A woman cutting across a field to take lunch to her husband, kicks out a nugget worth hundreds of dollars.

A Wells Fargo pair of scales weighed gold dust worth $55,000,000. Sugar cost $3 a pound, molasses $5 a gallon, flour $150 a pound, onions $1 a pound, sardines and lobsters $4 a can, candles 50c each; knives to pry nuggets from crevices $30 each. Contrasting prices when the panic hit about 1871 were: tenderloin steak 20c, fried ham 10c, liver 10c, ham and two eggs 15c, soup 5c, fried salmon 10c, rooms $1.00 per week.

(122) "Springfield" — 1850

Kelley's Saloon and peach orchard where the famous "Kelly" peaches returned 50c each.

(123) "Jackass Hill"

Scene of a very coarse gold — 100 sq. feet yielded $10,000. Mark Twain sat under the huge oak here to write the "Jumping Frog of Calaveras."

(124) "Roaring Camp"

Scene of the "Luck of Roaring Camp."

(125) "Melones" (Slugmullion)

One of the toughest mining towns in the state during 1849. Nuggets were called "Melones" by the Mexicans because of their size. The Melones Mine is still operating as a very modern and profitable venture.

Vignoli's store, former gambling den at Melones, one of the toughest mining towns in the state in 1849.
— Stockton Chamber of Commerce

(126) "Carson Hill"

One of the world's largest nuggets was found here, weighing 214 pounds troy, worth $44,000. In 1850, a man named Hance, following a straying mule, saw yellow metal on a quartz outcropping. With a rock he knocked off a chunk which proved to contain 14 pounds of gold. The rich vein produced $110,000 in gold from a single blast of dynamite. It became the Morgan Mine.

(127) "Heavytree Hill"

(128) "Sumpson's Bar"

(129) "Wayne's Bar"

On the Stanislaus River is the locale for some of Bret Harte's stories.

(130) "Vallecito" (Little Valley)

A roaring mining camp in the early '50s.

(131) "Douglas Flat"

The old Post Office and the Gilleado Building still stand.

(132) "Murphy's"

Bret Harte's "A Night in Wingdam" was referring to the old "Sperry Hotel." The old bar where Joaquin Murieta was "called" by Sperry bears evidence of the gun fight with a bullet hole in the door. Rollens Store was owned by

a French Army officer who came to California to make his fortune. He sold out his little store for $15,000 and went back to fight the Franco-Prussion War. Here it was that Joaquin's brother was unjustly accused of stealing horses and hanged. This incident led to Joaquin Murieta's career of banditry.

A street view of Murphy's where Joaquin Murietta was "called" by Sperry at the "Sperry Hotel."
— California Department of Natural Resources

Sperry Hotel at Murphy's.
— Stockton Chamber of Commerce

(133) "Sheep Ranch"

Where Senator Hearst lived and worked, laying the foundation for the Hearst fortune. The Hearst Mine was near the old Hotel.

(134) "Angel's Camp"

Named after George Angel, an ex-soldier in 1849. On the Stanislaus River at this point Angel discovered gold, which led to the founding of the town. An old arrastre for crushing gold ore was operated by a water wheel. Bennager Raspberry, the namesake of Raspberry Lane, shot a ramrod from a jammed gun, which landed at the roots of a manzanita bush. Upon retrieving it, Raspberry found a quartz vein from which he took out $700 that afternoon; $2,000 the next day and $7,000 the third day; then worked the mine at a profit for months. Angel's Camp is the scene of Mark Twain's "Jumping Frog."

Angel's Camp. The Angels Mine is now dismantled
— Stockton Chamber of Commerce

(135) "Albany Flat"
A placer camp ruled by the sporting element.

(136) "Altaville or Cherokee Flat" 1857
Locale of Bret Harte's "To the Chiocene Skull."

(137) "Copperopolis"
A vision point for stage coaches once had a population of several thousand. Cobb Corner Hotel was the hangout for the bandit "Black Bart." The Ames Mine produced 20,000 tons of high grade copper ore per year which was sent to England.

(138) "Dogtown"
Is only a memory, although it was once a wild mining camp.

(139) "San Antone Cap"
Was a rich placer camp.

(140) "Fourth Crossing"
Bret Harte tried to be a miner here but had little success in panning. This was a relief point for the old stage coaches with its Herrick House for over-night stops.

(141) "Calveritas"
Rich placer camp with an old relic of a building which served as a saloon, bank and fandango hall.

(142) "Scratch Gulch"
Nothing remains but the name.

(143) "Brandy Flat"
A gold camp.

(144) "Mountain Ranch or Eldorado"
Notorious wild mining camp with rich placer and the tiniest full-fledged post office in the state.

(145) "Poverty Flat"
Old man Tollinsbee and daughter "Lily of Poverty Flat" lived here.

(146) "Whiskey Slide"
The name resulted from the effects of "Jack-ass Brandy." A few drinks and all the miners were sliding pell mell down a steep incline shouting their heads off. It is a thriving mining camp. Here $15,000 is supposed to be buried by a miner who was murdered in an unsuccessful attempt at robbery.

(147) "Jesus Maria" — 1850
A setting to delight the eye of an artist.

(148) "Happy Valley"
Here was born the first white child of Calaveras County — Jenny Boudin. Old tunnel leads to winery from Boudin's store.

(149) "Mokelumne Hill" or "Mok Hill"
A large and lively town among the 'diggins.' Joaquin Murietta enjoyed disguising himself, playing cards with the miners and discussing the best method of his capture. The rush to this locality was started by a negro fugitive from a nearby jail. As a joke he was given an outfit and directed to dig in nearby "Nigger Hills." After a few days he returned to camp laden with nuggets. The ground aroung "Stockton Hill" was so rich that claims were restricted to sixteen square feet. At a burial one of the miners volunteered to read the service. While he prayed, with one eye open, he saw a miner pick up a nugget out of the dirt thrown from the grave. The preacher shouted "Gold!—Hold on boys, postpone this funeral until we locate out claims." "Chinatown" boasted 1000 Orientals.

Remains of adobe buildings belonging to Chinese who figured in many mining centers.
— California Department of Natural Resources

(150) "Chile Junction and Chile Gulch"

The scene of the "Chilean War" when American miners objected to the importation of peons or slaves from Chile to work Dr. Concha's claims.

(151) "San Andreas" — 1859

Joaquin Murietta started his criminal rampage here. There were 21 men involved in the hanging of Joaquin's brother on a trumped-up charge and the flogging of Joaquin. All of the men met violent death, nineteen at the hand of Murietta, who left his trademark of vengeance on their foreheads.

(152) "Pioneer Cemetery"

Two miles west of San Andreas.

(153) "Double Springs" — 1850

The first courthouse of Calaveras County was built of aromatic oriental wood imported from China.

(154) "Milton"

Harbored the bandit Black Bart often at the old Tornado Hotel without the proprietor realizing it.

(155) "Jenny Lind"

"The Swedish Nightingale" sang here in the early days. The old Sinclair adobe store and the ruins of the Rosenberg Building still remain after ninety years. The gold was taken from the river nearby by dredging.

Jenny Lind: Rosenberg Building ruins, after 90 years.
— Stockton Chamber of Commerce

(156) "Lancha Plana" - "Flatboat"

The sister of Joaquin Murietta lived here. She was often visited by Joaquin after the murder of his Rosita. Sam Brown, the notorious bandit, made this a rendezvous. Gold dredging is still going on.

(157) "Boston House"

A combination store and inn on the old stage road to Jackson.

(158) "Big Bar"

Marks the location of the old Whale Boat Ferry of 1850. Replaced by a toll bridge in 1852.

(159) "Butte City"

Site of the Ginocchio store built of stone imported from China.

(160) Jackson "Bottileas"

This was an overnight stopping place for freighters hauling Drytown and Mokelumne. Their whiskey botties accumulated in such great numbers that the place was called "Bottileas" — later changed to "Jackson" in honor of Colonel Jackson. This was the home of the Marre family who owned the wholesale liquor house.

Two of the world's deepest mines, the "Argonaut", 6,600 feet, and the "Kennedy", 5,500 feet, are still operating in gold producing quartz, which has returned millions of dollars.

Jackson Jail at **Jackson.**
— Stockton Chamber of Commerce

(161)

Railroad Flat was old 65 years ago, but nothing remains but a store. It was named from the wooden rails used in the mine.

(162) "West Point"

Bret Harte loved this place and its surroundings even though he played in hard luck. His try at mining proved a failure, and most of his early stories brought nothing but rejection slips.

Kit Carson and hardy scouts camped here in 1845 and named the town. Louis Chicard sent the first mule train into this camp loaded with all sorts of merchandise. An ancient river bed produced pay dirt for a distance of 20 miles along its course. Some nuggets were valued from $5 to $90. The channel produced many millions in gold. Nearby is the granite belt of the East Mother Lode and there stand magnificent pines through which flow many auriferous creeks with their lively trout.

(163) "Volcano"

Now a phantom city, but its population numbered thousands in 1854. Some of its landmarks remain and members of the old pioneer family of Bonneau have kept the old store functioning.

The old church where Thomas Starr King used to preach still stands. The neighboring mines produced $90,000,000 in gold.

Volcano, in 1854 having a population of thousands, now a picturesque "Ghost Town."
— Stockton Chamber of Commerce

(164) "Pine Grove"

Nine miles east of Jackson on the Alpine Highway. Once a populous center with numerous saloons, dance halls and churches. Now it is deserted.

(165) "Tovey Monument"

It is located three and a half miles west of Martell at Morrow Grade, where Black Bart used to hold up stages. Mike Tovey, Wells Fargo messenger, was killed, and DeWitt Clinton Radcliffe, stage driver, injured at this place in 1893. The main artery to the Mother Lode ran between Sacramento and Sonora via Q Ranch near Ione, Jackson, Mok Hill, Angelo and Columbia. Over $265,000,000 in gold bullion was carried over this artery in the early days.

(166) "Sutter Creek"

Named after General Sutter who built the first sawmill on Sutter Creek. Much of the atmosphere of early days remains. The Union mine produced for Leland Stanford the money for investment in the Central Pacific Railroad, from which he made millions.

Meta-andesite fieldstone building on hill east of Sutter Creek, named after General Sutter who built the first sawmill on Sutter Creek.
— California Department of Natural Resources

Winery remains on Volcano road east of Sutter Creek.
— California Department of Natural Resources

Looking south at Sutter Creek.
— Stockton Chamber of Commerce

(167) "Amador City"

This was a lively Camp in '49. Abandoned hotels and stores still stand as monuments to a famed city. Site of the famous Keystone Mine, a brick office building which still stands.

Well preserved stone building in Amador City.
— California Department of Natural Resources

(168) "Drytown"

Founded in 1848, it is the oldest town in Amador County. It contained twenty-six saloons, so wasn't very dry, as its name implies.

54

53

It was located on Dry Creek. The nearby gulches were very rich, it being not unusual to wash out $100 in gold from a single pan.

A brick store, converted into a gasoline station in **Drytown,** where one can find in the surrounding hills a score of abandoned mines, whose names were once a household word.

— California Department of Natural Resources

A clay pit in Amador County, locale of the famous Keystone Mine.

— California Department of Natural Resources

(169) "Forest Home"
Favorite stopping place for stage coaches. It boasted a winery, saloon, dancehall and store.

(170) "Michigan Bar"
Chinese miners bought the old Heath Store, destroyed what was left of it, and carried on a very thorough search for gold.

(171) "Oleta"
Means "Old Home Springs", but it was called Fiddletown in the early days because of its numerous Missourian fiddlers. Bret Harte's "An Episode of Fiddletown", probably used the St. Charles Hotel with its lounging red-shirted miners as the Fiddletown Hotel. Chinatown, on the outskirts, stretched for over a mile along the road, consisting of wooden shacks, stores, gambling dens and tents.

(172) "Shingle Springs"
Built of lumber brought from around the

horn. It has been preserved and is now "Mary's Inn". It dates from 1850

(173) "El Dorado or Mud Springs"
Some of the old buildings remain. One brick building, which was owned by Charles Jackson, now has a tree growing on it. In a single night, $10,000 in nuggets was weighed out.

Rhyolite tuff and brick buildings in **El Dorado,** which boasts a larger number of 1850 buildings than many other Mother Lode towns.

— California Department of Natural Resources

Rear of two-story stone building shown above.

— California Department of Natural Resources

Tuff-faced meta-andesite building still completely intact in **El Dorado.**

— California Department of Natural Resources

(174) "Diamond Springs"

In 1860, Edmund Cooper located a cabin here, took out a nice stake and buried it under an oak tree. He left for other diggings, not wanting to carry his wealth with him. Years later he returned and found a railroad had cut across his claim where the oak tree and gold was. Cooper, in 1933, was 118 years old, a resident of the County Hospital. The old graveyard bears stones dating as early as 1851.

(175) "Placerville"

Was known in 1848 as "Dry Diggins' " and later became "Hangtown" because of the hanging of a white man and two Mexicans from an oak tree on "Main Street". In front of the Carey Hotel, Hank Monk, the ace of stage drivers, changed horses. Horace Greeley rode into town with him. Mark Twain's story "Roughing It" used Monk as a character. The old Wilcox Warehouse was the depot for the stage line from Placerville to Lake Tahoe over the toll road. John Studebaker failed as miner, but did a flourishing business as a wheelwright and blacksmith. Mark Hopkins was a grocery man.

(176) "Lost Indian Ledge" California

General Fremont took a southerly route on one of his exploration trips which took him from the Antelope Valley to the San Joaquin Valley. Tributary to this route, as it entered the mountains, a band of Indians held sway. One of their number, a young buck, committing a crime, is made to pay for it by being tossed over a high cliff and left to die. Though badly injured, the Indian survives the ordeal, and is found by his Indian loved one, who, with care and knowledge of herbs, cotton plants and spruce, brings him back to health. They ranged a beautiful country with plenty of game and water. During one of the excursions for food, the Indian discovered a ledge of very rich gold ore. The ore was so rich that crude pounding and crushing methods, together with washing, produced a considerable sum of gold, which enabled the young Indians to live in security from their tribe.

Look for a long canyon bearing northwesterly toward the San Joaquin Valley. In the vicinity of Fremont's entrance to this canyon is the Gold Ledge.

(177) "Bodie" — California

Grandfather Honeywell of Bridgeport, California, established a sawmill on Buckeye Creek to supply lumber for the booming gold camp of Bodie. Old relics of the lumbering operation lie around where discarded. The site of this old lumbering operation has considerable scenic allure. Buckeye Canyon is extremely well watered and, therefore, appreciated by beautiful patches of colorful flowers forming a contrast to the surrounding dense greens. Honeywell had considerable choice as to open ground for placer mining and for some reason chose a large part of a section lying north of Mono Lake, about five miles. The gravel is sprinkled with fine gold and in the draws nuggets were found. This land hasn't been mined and lies awaiting some method for the extraction of fine gold. To the east and south of Bodie lies an area which has produced gold for the Indians, which they used for trading purposes. No record was kept as to the amount but some of the later generation of Indians having heard stories of the gold nuggets, go into the territory in search of the source. Bodie and the settlement of Benton Station were trading posts for the Indians.

(178) "Bridgeport" — California

Many adventurers and gold seekers enlisted in the Army in order to gain some knowledge of the far west and have expenses paid while doing so. During the enlistment, a soldier could find a little time for prospecting. If a certain section looked good, he could return to it after enlistment. About five miles west of Bridgeport a monument has been erected designating the campsite of General Fremont. Old records tell of some of the soldiers finding gold in the vicinity. Recent years have furnished evidence that this was so. Two miners came into Bridgeport at regular intervals with gold dust and nuggets to trade for supplies. As is usual in such cases, they were followed on their return trip, but after turning off the main road and entering the timber hills to the north, they were lost to their trailers. Recently the writer entered into conversation with an old gentleman in Los Angeles, a former attorney, who went in for part-time prospecting. He helped locate some claims northwest of Bridgeport about six or seven miles that showed very high value in telluride, a silver-gold combination.

(179) Kernville — California

Many rich pockets of gold have been discovered in the area tributary to the Kern River, around such settlements as Isabella, Claraville and Havilah. Some very beautiful speciments have survived the mining of this district. The ore was extremely rich and as a result the area proved very fertile for the origination of tales of lost mines.

Some good mines are producing not far from Kernville, so perhaps there was some justification for a few of the many tales of lost gold. One of interest deals with a trip into the hills into the vicinity of Havilah. Sam Holmes, an old resident of Los Angeles, related his experience in making a trip into the district about 1890. His party took a train to Caliente, stage to Havilah and a hired buckboard and team to make the journey into the hills. A mining man, hearing of some claims that were open for entry, decided to investigate. These were reached and found to be located in a very scenic area, well-watered and timbered with large yellow pines. There were five claims in the group which were resurveyed and staked. Then samples of ore were taken and Sam was told that it looked fine. He was advised to take some adjoining claims both for mineral and timber. This he neglected to do. The return trip was made to the stage line. Sam rode the high seat of the thoroughbrace coach with the driver and had his curiosity aroused as to the package acting as a foot rest for him. He inquired of the driver, and was told that he would find out when they reached the railroad. It proved to be a couple of gold bars worth $1500. Two miners were averaging that much each, per month, back in the Kern River area. The stage agent weighed the gold, gave them a receipt and shipped it, giving them credit on the books. Sam's friend dropped dead on the street in Los Angeles a few weeks later and Sam Holmes is now dead. The exact location of the claims is not known, but Havilah is the closest settlement.

(180) Santa Fe — New Mexico

Coronado explored this area about 1540 and Don Juan de Onate established the Colony of Santa Fe in 1595. The Indians were used to working the mines and received harsh and cruel treatment under the Spaniards. Many valuable mines were discovered and worked for years, but because of the brutal treatment and slavery of the Indians, an insurrection and massacre resulted in which the greatest portion of the Spanish population was killed. Those surviving, retreated southward under Governor Otermin and founded the settlement of Paso del Norte, 320 miles from Santa Fe. The massacre took place in 1680 and at that time many mine entrances were concealed by the Indians and locations lost completely. Commerce was carried on between Santa Fe and Chihuahua, Mexico, by such men as Josiah Gregg and his parties, which proved very profitable.

In 1838, a year's adventure on the part of twenty-three Americans netted them $150,000 in specie and bullion.

(181) El Paso

A trade route was established between Santa Fe and El Paso, a distance of 320 miles, and then extended to Chihuahua, a total of 550 miles, requiring 40 days' travel. J. Gregg made this journey about 1838. Trade was profitable in spite of the usury on the part of a few Mexican officials, and, therefore, much bullion was transported back to Santa Fe. Between Robledo and El Paso are many ruins of settlements destroyed by Apaches. The Sierras, separating the Rio Del Norte and the Rio Pecos, are visible to the east of the trail. Sierra Blanca and Los Organos are two of the Apache strongholds. The river was forded about 6 miles above El Paso and from there to Chihuahua the journey must have been anything but a pleasure. From El Paso to Los Medanos, sand stretched endlessly with widths up to six miles and for a distance of 60 miles the only springs were fetid. A fort at Carrizal and beyond an Ojo Caliente (Warm Spring) and then the valley of Encinillas which seemed very fertile.

If one should travel this old trail and live in the past again keeping company with some of the hardy transporters of specie, gold or silver, or trailing a band of marauding Apaches, he would be richly rewarded in the knowledge of their hiding places for treasures.

(182) Jesus Maria

A mining town in the department of Chihuahua, was very active in the year 1935. Situated in the heart of the Cordilleras, 150 miles west of Chihuahua, the territory on its approach was extremely rough, with steep trails along the border of precipices and the use of many switchbacks to reach the mine floor from a point high in the mountain above. At this time silver was being mined at depths as great as 800 feet with mule power and windlass. The specie was packed out on mules, each animal carrying a pair of sacks worth up $2000. To avoid friction and injury to the animals, the specie was packed in sacks of raw beef hide, which, upon shrinking, compressed the contents.

(183) Chihuahua

Was founded in 1691. In 1839, it had 10,-000 souls. The Santa Eulalia Mines nearby contributed much wealth, the Church receiving $1,000,000. The rich ore was sent to the smelting furnaces, melted down and extracted from the virgin fossil. If inferior, the ore is amalgamated. The Moliendas (Arrastres) are the crushers. A circular cistern 20 to 30 feet in diameter is dug in the earth and the bottom and sides lined with hewn stone of hardest quality. Transversely, through an upright post which turns upon its axis, passes a shaft of wood at each end of which is attached by chords, two grinding stones with smooth, flat surfaces, which are dragged by mules fastened to the shafts.

The ore is first pounded into small pieces and ground with the addition of water and quicksilver to the resulting mortar. Some muriates, sulphates and other chemicals are added to facilitate amalgamation. The compound is washed in a stone tub with a revolving upright studded with pegs and a constant stream of water. The light matter is carried off at the top and amalgamated metals sink to the bottom. The quicksilver is pressed out and silver submitted to a burning process by which the balance of mercury is expelled. The silver coming from the furnace contains about 10 to 30 per cent gold. The gold, while in a liquid state, settles to the bottom, but still contains considerable silver. This compound is called "oroche". The silver is moulded into barras (ingots) weighing 50 to 80 pounds and worth $1000 to $2000. They are stamped with weight and character and constitute a species of currency. People of wealth lay up their funds in ingots and some of the cellars of the "Ricos" present the appearance of stored fuel.

(184)

There are many famous crossings of trail and river. Here the hunter for specie, bullion and gold nuggets should make investigation. About 4 miles south of the Cimarron at Upper Spring in Oklahoma, the early trail riders had a delightful camping place. Grapes, wild plums, currants and berries grew in abundance and game was plentiful. At El Vado de Piedras is a ford which is eighty miles from the source of the Canadian River. The north fork of the Canadian was crossed at Eufaula, McIntosh County, Oklahoma. At Little River is Camp

Holmes, with a Spring, and nearby, plenty of deer, turkey, partridge and grouse. The celebrated "Caches" on the Arkansas River afford

an inspection of the method used in burying treasure, usually a hole was dug at a point of some elevation to secure it from inundations, the sides lined with sticks and grass, and cut sod carefully replaced. The owner of the treasure had to be very careful to map it if he wanted to return to it, as the bottomland with all its sod looked alike. Sudden attacks by Indians forced the early traders to bury their treasures without delay. The values were high in many cases as the individual operators gathered at a central point and then trailed back to the east with their goods in large bands for safety.

(185) Laffite and Pirate Booty

Two brothers and Pierre Laffite enter the story in New Orleans about 1803. They ran a blacksmith shop but had most of the work done by slaves. Their real business seemed to be over on Barataria (Smugglers Retreat). The island of Barataria or Grand Terre lies between the Gulf and Barataria Bay in an area which is still the haunt of wild things and transversed mostly by the pirogue of the pot hunter or the negro moss gatherer. The men who plied their piratical trade in this region were a motley crew — Portugese, French, Italian and Malay. The Baratarians were privateers — licensed pirates. Armed with letters of marque against Spanish shipping and authorized by France and the United States to prey on English commerce. With plunder in hand and a market at their door, Jean Laffite in New Orleans became their agent. From agent to chieftain was only a step. His power grew until arrested by the Governor of Louisiana, when he engaged the two best lawyers in the State at a fee of $20,000 each. One of the lawyers was invited to Barataria to receive his fee. He was feasted for a week, then returned to the mainland in a superb yawl laden with boxes of Spanish gold and silver. When the United States declared war on England in 1812 the British prepared to lay siege to New Orleans.

They enlisted the aid of Laffite with promise of large rewards, but he, in turn, made their plans known to the United States officers and offered his services. As a reply to these friendly advances, the United States Navy attacked the privateer stronghold and captured a large quantity of booty. Laffite was away at the time and most of his men escaped and fortified themselves on Last Island. When Andrew Jackson reached New Orleans to fight the British, he decided the Baratarians were men of his own mettle and put Laffite and his captains in charge of two important batteries. The battle was won and President Madison issued full pardon for all Baratarians who took part in the battle. It was in 1816 that Jean and Pierre undertook to make a new Barataria at Galveston Island. Jean was called the "Lord of Galveston Island." He was at this time about forty years old and is described as being exceedingly handsome, even noble in appearance, with magnetism, charm and suavity. He seldom smiled, but cultivated in a rare manner the art of being agreeable. Generally, he went unarmed, but when aroused, he was a desperate man indeed, being an unerring shot.

Laffite usually received a "royal fifth" of all rich hauls. Jim Campbell, one of Laffite's most famous captains, told of one cruise of six weeks duration when he captured cargoes valued at $200,000. At another time he captured 308 slaves. They were valued at $1 per pound. Laffite abandoned Galveston at the request of Lieutenant Kearney in command of a United States Man-of-War. Laffite is supposed to have carried on his operations from the mouth of the Lavac River which consisted of treacherous sand bars through which meandered a narrow channel. When pursued by a United States revenue cutter, which had discovered his "Pride" cruising for a prize, Laffite was forced to enter the Lavaca River and run it to a point where he was bottled up and forced to abandon ship. Nosing into the bank with his sinking ship, Laffite and his men divided the treasure and separated. Only two remained with him and when they went ashore, they carried a chest containing a million dollars worth of gold and jewels. It could not be carried very far, so it was buried in a salt grass flat about a quarter of a mile east of the Lavaca River. A Jacob's staff was placed over the chest, bearing taken on two mottes of trees within sight, then the staff was left exposed only a foot. After several days of hardships, they were cared for at a settler's cabin. Here Laffite told his two companions that he was going among the Indians on Red Rover to the north and that if they found the chest unmoved after three years they might have it. Months later, one of these men, while dying in a saloon in New Orleans confided his secret to an Irish bartender. The other married and had two sons, and when they were mature, told them facts that corroborated the bartender's story.

The two sons searched for Laffite's chest many years ago. A rancher named Hill acquired the land down the Lavaca River and stocked it with horses. One day a negro took some mares down to a flat along the river, and wishing to stake his saddle horse, kicked around in the salt grass for a piece of wood. He barked his shin on something hard and discovered a brass rod, fast in the earth with a socket on the end, the very thing for a stake pin. The negro took a nap and was aroused when his horse became entangled in the rope and discovered the stake almost out of the ground. He took the rod to the ranch and Hill knew it was a Jacob's staff and what it meant near the Lavaca River. He at once ordered the negro to lead him to the spot. Perhaps the negro was lazy and did not want to go so far; perhaps he was so ignorant of woodcraft that he could not find the place. Anyway, he did not lead Hill to the spot. Hill tried to trail the negro and his horse, and continued the search over a long period, but the whole country was covered with a thick grass tracked over everywhere by horses.

Then, quite recently, J. C. Wise from San Antonio, and companions, started a search, first looking for the old ship in the Lavaca River. This they found and sighted the two Mottes; then using a mineral machine, they began the search. For five days, in rain and mud, they searched. Wise insists that the chest is still there, and some day he will go back after it.

(186)

The mouth of the Colorado River is as good a place to look for pirate treasure as the Lavaca . . . William Selkirk of New York befriended a very sick sailor named Robinson. The sailor was about to die so told his benefactor his life's story and gave him a map. The sailor had been a pirate under one of Laffite's captains. Once while they were cruising for a prize in the Gulf of Mexico, morning showed them that they were very near an armed Spanish ship. The pirates' brig ran for shore and slipped over a bar into the mouth of the Colorado River. The Captain decided to remain here until dark and then take a chance on escaping. In the meantime, he gave orders that the ship's treasure chest be buried ashore. Robinson and two sailors did the job, giving the captain a detailed description of the spot, henceforth called Gold Point.

The Spanish ship came in close enough to open fire. The captain was killed as were the two sailors. Thus he alone had the exact knowledge of the treasure. Escaping the Spaniard, the next several years were spent sailing over the world wherever the winds of fortune called, then with the chest still unrecovered, he was dying, an object of charity, in New York City.

William Selkirk was much impressed with the sailor's story and had faith enough in it to pull up stakes for Texas. From the Mexican government he purchased 6000 acres of land at the mouth of the Colorado River — including Gold Point. The Selkirks have kept the land and paid taxes on it all these years.

They hired a negro by the name of George Ellis to watch certain marshy tracts and keep off squatters or treasure hunters. Some men approached him some time ago with the proposition of hiring him to aid in the search for a chest. George figured that by working with the intruders he could watch them and protect the Selkirk interests. After they had dug a while, George felt his spade scrape against metal that he knew was the lid of the box. He sat down on the box lid and complained to his boss that he was too full of misery to work any more and that he was going to quit. Also, he suggested that the others were not going to find anything and might as well quit too. Much to his surprise, his employers did quit on the spot and set out toward Bay City. Had it not been for Negro ha'nts, George would have returned under cover of darkness to finish the job, but instead, it was after sunup when he returned. To his dismay he found a rectanglar hole in the earth where the chest had been lifted out during the night. One of the hunters, presumably, had struck it with his spade about the same time that George made his discovery. The negro's getting out of the way was just what the white men wanted.

(187)
The mouth of the Neches River, according to a chart owned by a man named Marion Meredith, is the site of a tree with a chain about it. The chart tells of a ship cabled to the tree, the planting of a fine "wad" of loot nearby; the shelling and sinking of the ship by a Spanish galleon with only one man surviving, the man handed down the chart. This loot hasn't been found.

(188) Kokoweef — California
Rising from a plateau which is about 5000 feet above sea level, Kohoweef continues on up another 1000 feet to attain the dignity of a landmark. Situated in the Ivanpah Mountains where the atmosphere is very clear, and elevated as it is, one can see hundreds of miles north, east, south and west. The Indians must have delighted in it as a point of vantage. They apparently visited the region as one story relates how two Indian boys — brothers — were climbing the mountain when they discovered a small opening. Investigation showed a large cave which continued to the interior for some distance, then a drop-off which held a pool of water. They tested the depth and extent by throwing rocks into the pool. One of the boys noticed an unusual weight to one piece so put it in his pocket for investigation.

With good light they saw that they had a piece of rock well-studded with gold. Returning later, they filled their pockets with nuggets, then decided to have a swim in the pool. One of the boys, while diving, hit his head on a projecting rock. He was pulled out of the pool by his brother, but too late to be revived. Indian superstition wouldn't allow any of the Indians to re-enter the cave to get more gold and the entrance was closed. Years later, a prospector, John DeLano, while swinging his pole pick, was aware of a hollow sound. Digging showed an entrance to Kokoweef and he investigated. There wasn't any water in the cave and he found no nuggets. At the present time a mining company is developing a lead zinc mine on the mountain. They will drive a tunnel right through its entire width.

(189) The Mormon Battalion 1846 - 1847
A story of one of the most amazing infantry marches in history. Colonel Philip St. George Cook and his Mormon soldiers broke the first wagon trail through a rugged and hostile desert from Santa Fe to San Diego, a distance of 1100 miles. This was just prior to the discovery of gold in California in 1849. Their route was used by many gold seekers in the years to follow. The Mormons were not wanted in Missouri and were denied admittance to Arkansas. It seemed best to emigrate to the far west. This group decided to enlist in the army and gain their objective with government aid. The government, on the other hand, hoped to use them to fight the Mexicans in California and also the colonizers. General Kearney hurried ahead with his dragoons on horseback and ordered Cooke to open a wagon road through the southwest to California. The start was made at Council Bluffs, Iowa, on July 21, 1846. It took until October 10 to reach Santa Fe. Kit Carson was guide for Kearney and had advised against taking the wagons with the first contingent so they were assigned to Cooke. Colonel Cooke, with 397 troops and 60 days' supplies of flour, coffee and salt, 30 days' rations of salt pork, and 20 days' supplies of soap, and 26 creaking wagons, left Santa Fe on October 19th.

Guides to aid the party were Pauline Weaver, Antoine Leroux and Charbonneau. They proved little assistance. Mules gave out and died and men became sick. November 9, fifty-eight of them were ordered back to Santa Fe. The Rio Grande was followed to a point near the silver mines of New Mexico. Deer and turkeys supplemented the meager rations at this point. On November 13, the Rio Grande was left behind near the present town of Rincon, Dona Ana County. Their next move was to the Mimbres River; then the Ojo de Vaca. At this point the old Spanish Trail between Janos Chihuahua and the Santa Rita copper mines was followed a short distance to the south, then left this trail to cross Whitmore

Pass and follow Animas Creek. Grizzly bears were seen here, black tail deer and antelope. Manuelito, the Apache Chieftain, visited the camp near Guadalupe Pass. On December 2, 1846, the weary Mormons reached the old Rancho of San Bernardino in the valley of that name. Large adobe buildings with high walls for defense were intact but the ranch was deserted. Wild cattle from former Spanish herds supported the Apaches of that district and delicious fresh meat of wild oxen allowed the hungry Mormons a welcome diversion of fare. The San Pedro river was reached on December 9. Here wild bulls attacked the column of marching men, wounding several and killing a number of mules. Lieutenant Stoneman (later Governor of California) shot himself accidentally in the hand. The men caught trout three feet long in the San Pedro.

In Tucson, Cooke found inhabitants very friendly. He had expected trouble from the Mexican Commander of the Presidio. Between Tucson and the Gila River the marches were made at night because of the lack of water and the heat. When the Pima villages on the Gila were reached, the soldiers found about 2000 friendly and cheerful Indians with produce for sale — even water-melons. Having trouble along the Gila with sand, Cooke decided to float the heaviest supplies downstream. Stoneman was put in command but the supplies had to be cached. However, the floats aided in ferrying the battalion over the Colorado which was reached January 9th near Algodones, Baja, California.

Then to El Alamo, a waterhole; Pozo Hondo, where they received word of the battle of San Pascual; Carrizo Creek with running water and relief from real thirst and bleeding feet; Vallecito and San Felipe where a road was chiseled out of rocks; Warner's Ranch January 21, 1847; then Temecula, heading for Los Angeles, turned back, and on January 29, 1847, reached San Diego.

(190) Death Valley Scotty's Mine

During the very active mining period which started just after the depression of 1907, the writer was working on a survey that was being made for the railroad, running from Wagon Tire Mountain to Lakeview, Oregon. One of the Scott boys was hired as a teamster at Lakeview, this being the home place for the Scott family. The brother, in accounting for a very pronounced limp, told of the invitation he had received from Scotty to visit him in Death Valley and that he would be shown the gold mine that was such a mystery to the family and the public in general. The brothers met in Death Valley; packed enough grub to do for several days, provided for water, and set out with a pack outfit. The first night out they made dry camp, wood being as scarce as water. Scotty said his brother seemed very pleased to hear of the family and they talked well into the night. Sometime before daylight Scott

was awakened by the mules who were causing a disturbance. He spoke to his brother but received no reply, and got up to investigate. He started out to circle the camp and had gone but a short distance when a shot rang out and he felt himself the recipient of the bullet. When Scotty arrived, he explained that he had heard a noise and got up to see what it was but he didn't see anyone. The bullet had made a nasty wound in the leg necessitating a return to headquarters with all speed. Scott said he didn't know who shot him or why. Rather mildly, however, he surmised that his brother didn't have much of mine.

Scotty has been seen with pack animals and Indians far to the south of his present home in the vicinity of Quail Spring where some rich float has been picked up in recent years. No gold mines have been developed in this area. However, there has been much activity just to the south and east of Quail Spring but the mining has been for manganese.

(191) "The National Mine" 1908 — Nevada

Situated in the Santa Rosa Mountains at an elevation of 7000 feet, about 20 miles southeast from Old Fort McDermit, with ore running as high as twenty-five dollars a pound (not a ton) which is good enough. It caused intense excitement. The Stall brothers took out four million dollars in three years. The ore was a tellrude and being so rich in gold made highgrading very tempting. Miners were searched when they left the drifts; the entrance was guarded by men with rifles; floodlights were used at night; yet in spite of all precautions there was considerable highgrading. The lessors did as little tunnel work as they could, preferring to follow the "blue gouge" which was the associated material, by working in a kneeling position and excavating just enough to allow them to follow the veins. Thousands of feet of drifts were made in this fashion, saving the miners a great deal of hard work. Those going in for highgrading usually held out a small portion of the richest ore each day by secreting it in the main tunnel, usually on top of the supporting cross timbers. After an accumulation, it was gathered up and taken to the outside. The aid of the shift boss was necessary as he wasn't searched and in some cases even better paid bosses entered into the game.

A few years ago the writer was trying to revive interest in a group of claims adjacent to the National. A tunnel had been driven 1800 feet along the south line of the National property. Lack of money and water caused a halt. Jim McAllister, one of the owners showed his wife a piece of ore that came from his claim, not much larger than a chicken egg but worth several hundred dollars. Jim McAllister is dead; a confused record remains, and the tunnels have caved in or contain water. National City was laid out near two bountiful springs in a mountain meadow. Nothing remains but piles of broken whiskey bottles and the iron cook stove which is about ten feet long and four feet wide with a large part of the top a smooth heavy plate ideal for flapjacks or steaks. As a guess as to the method for carrying such a huge piece of metal from the valley floor to the top of the mountain, the picture presents itself of a heavy ore wagon built high off the ground with heavy hickory wheels bound with thick metal hoops and drawn by six to eight horse teams.

Seated around a camp fire on the site of National City, one old timer was proudly telling us of his exploits as a high-grader. He had participated in the rich awards of the National Mine and many others, but the one he couldn't forget was over in Colorado where he had worked for many months and with the aid of others, accumulated a very sizeable pile of rich, free milling gold ore. With the knowledge of the mine manager, the loot was buried under his bunk house. About the time that resignations were in order, the next mine victim having been chosen, a fire of large proportions set in, engulfing the manager's bunk house as well as many others. It wouldn't do to be seen hauling out ore while rebuilding of the camp was going on. As far as the old timer knows, that rich ore is still there. He returned after the camp was rebuilt but couldn't get to the ore.

The two large springs at the site of old National City contribute much of the water that flows off the Santa Rosa Mountains into Paradise Valley. As the water accumulated to form ripples and pools, large trout are to be seen and caught. They are extremely hungry as evidenced when a large fellow is caught on a worm and due to haste in hooking him a poor job is done. He will drop off the hook, flop back into the pool, swim madly about in search of the tasty worm he remembers going after, then rushes the next worm that is thrown to him with such force that he almost lands himself.

(192) "One of the Lost Mines of the High Sierras, California"

At the north end of Mono Lake, a former miner by the name of Shepard holds forth. He was injured in a mine so doesn't get around as well as he used to, but his mining days are not over. About 300 yards from his house, he has some placer ground which to him is as good as a bank. When he feels the need of a good steak, a chunk of ham or bacon, he sifts a few nuggets out of the gravel and takes them to the store. Things like milk and butter, he can pick up at a ranch on the road to town. Vegetables, fruit, chickens and eggs he gathers right at the home front. Trout from very cold streams entering the lake are but a short distance away and in the mountains to the north,

deer and sage hens hold sway. The High Sierras start elevating themselves close to his cabin and Shepard will stand at the entrance and point to a contact of contrasting colors about 1000 feet up the mountain.

Just at the edge of the red area is a covered up shaft and its rich vein. The old miner who found the gold was taken ill and had to go to Carson City. He told Shepard that if he didn't get back he would let him know exactly the spot where he dug the shaft. One night in a dream, the old miner and Shepard were sitting at a table with a crude map before them with markings on it. The old fellow seemed excited and was earnestly trying to impress on Shepard the need of digging in a certain spot. A few days after this dream, Shepard received word that his miner friend had passed away. The shaft has not been located.

California (193) "The Doodle Bug Mine"

Near Goodsprings, some loveable old characters have put in the best years of their lives as miners. All their hopes of gain seems to have centered here. This group consised of Mars, Sexton, Tabor, Myton, Loop, Johnson, Burkett and Gill. Mars, Sexton, Tabor are dead. Johnnie Myton has retired, selling his Morning Star claims for $350,000. Levi Burkett lives alone in his neat, well-kept little cabin near his mine which had good ore but

also plenty of water, which presented the problem of how to get rid of it. John (Cap) Loop was recently thrown through the top of his auto as it was being driven by a soldier who had been given a ride and who offered to relieve Cap at the wheel. After several months in the hospital, Cap is back on the job, badly bent, but game as ever. He said he never did find out what became of the soldier. Cap and Joe Johnson pal together and hope their ship will come in, when they convince someone that their present claims are worth $100,000 or more. Some of them have holes, but one is still in its natural state. That's the Doodle Bug Claim. Cap became quite interested in geophysics as a possible means of locating ore bodies, so he bought himself a real outfit. He and Joe took it out one day to explore a likely looking outcrop. Cap had the earphones on when a signal came through with such strength that it just about knocked his head off. He called Joe over and his experience was the same. Cap said the response was too great, that something must have gone wrong, and after much discussion they decided that the spirit of their old friend Mars was telling them to go away and let him rest. No work has been done on the Doodle Bug Claim, but then the old fellows are in their seventies and hope the other fellow will do the digging.

(194) "The Indian Gulch Mine" California

The present owner of the Old Standard Mine No. 1 in the Ivanpah Range was prospecting a few years ago in the country south of the range. Some odd looking rock was found which seemed to be lava, being very porous but having some weight. Fracturing showed a dark colored metal which proved to be copper. A fair sample was assayed and the Hollywood assayer showed intense interest and offered to finance the further exploring. Riley Bembry, the man who found it, and a friend went back to search for the material, but had car trouble and ran out of water and barely got away with their lives. The find was of no special interest to the assayer, as he figured that if a volcano put out a flow material containing metallic copper, it could also contain gold and that such a combination lying there ready to be scooped up would have considerable value. The deposit remains as it was first seen by Bembry so far as he knows.

(195) "The Lost Mormom Mine" California

Near Government Holes in the Midhills Mining District, Bert Smith, a veteran of the last World War, rules the domain. He runs a few head of cattle, some goats, and prospects his seventeen claims. A few years ago, he was cutting across country and had his horse stop suddenly with a startled snort. The country all looked alike and Smith decided that possibly a rattler had frightened the horse, so he dug in the spurs. The response on the part of the horse was a leap which landed them into a pile of brush, and to Smith's dismay, a mine shaft. Horse and rider went down about twenty feet and the horse, in its struggle, kicked the daylights out of Smith. After a few weeks in bed, Smith recovered and went back to investigate. He cleared the hole of debris and found a well-defined vein of quartz which assayed forty dollars in gold. Later investigation convinced Smith that he had fallen into the shaft dug by Mormons several years before who had marked it by four juniper stumps situated in such a way that at the intersection of lines run from each burned juniper the owner could locate his covered shaft when he returned. He had gone to Barstow, reported

his discovery and then disappeared. The property has been leased, but little work done and its true value not determined.

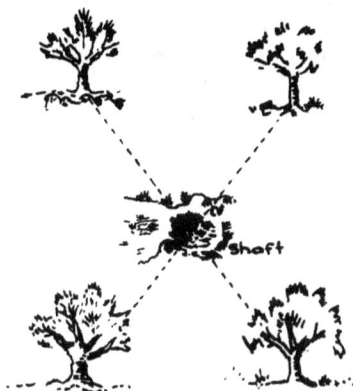

(196) "The Lost Mine of the Bullions" California

Ed Shaw has lived in the vicinity of Bullion Mountains for many years. The country to the south and west has produced placer in quantities sufficient to keep many a dry washer

Ed Shaw on a prospecting trip.

busy. Then at Camp Rock and at the Maumee Placer, heavy equipment was installed for the extraction of gold. This area is at the base of the Bullion Range, which, according to Shaw, was visited by father and son on a prospecting trip. They assumed that the gold in the gravels of the valley came from the higher country to the east.

They outfitted a burro which they hired at Daggett and set out for the mountain on foot. Their route took them past Shaws Camp near the Bessemer Mountain and their destination about a day's walk beyond. After a few days of prospecting, they uncovered a vein that they judged would run about $100 to the ton in gold, and its width showed permanence. They were going to stake their claim the next day, but found that the burro had pulled up its tethering stake and wandered off. They left their equipment in camp and went in search of the animal. Its tracks showed that it had headed for Daggett and they supposed they could catch up with it but they didn't reckon with the heat, lack of water, and distance. They reached Daggett half dead, having walked better than fifty miles with two drinks between them. They were so disgusted that they took the train for home which was in Pasadena, deciding to return later. This they did, and found that a cloudburst had destroyed all evidence of their camp and the ledge of gold ore.

(197) "The Mine of the Iron Door" California

Table Mountain lies just southeast of the Gold Valley Ranch, which is on the flank of the Providence Range. Bert Smith ranges this country, having decided that he would take care of that end of the mesa while Bob Holliman must stay at the other end. At least ten miles separated their two cabins. A dispute over water brought about a display of firearms but no shooting; just threats as to what would happen if either one trespassed in the future. Old Bob Holliman's knees don't bend very well and Bert's balance is bad, so they don't get about much now. One day Bert was looking for section corners and decided that one of them was located far up the slope of Table Mountain. He drove the pickup to the base of the mountain, then went the rest of the way on foot. The section corner was located and Smith decided the easiest way back was to slide down. All went well until a tearing sound convinced him that a good portion of his pants had parted. Going back to investigate, he found that it was the upper part of an iron door in place at the entrance of a tunnel. Debris from the mountain had almost completely covered it. Smith said he hadn't developed the property but that the ore which he later found at the base of the mountain showed very good value.

(198) "McGee Creek Gold" California

Two Indians were running sheep in the High Sierras at the head of McGee Creek. The Creek is in a narrow valley for about four miles, then tumbles down precipitous cliffs for hundreds of feet. The trail leading from the valley to the meadows above must gain elevation by the use of switch backs. Each winter, the rain and snow will damage the trail, making it necessary to blast new routes. This work is usually done in the spring, so between blasting and erosion, much of the material is dislodged and carried to the valley below. George Brown, an Indian packer, who lives near Bishop, was the source of information regarding the gold ore picked up by the sheep herders. They were coming down from the high country for the last time that season. The pack on one of the mules had shifted and in order to save a job of retying, one of the men picked up a large heavy rock to act as a balance. Reaching the low country, they decided to tighten the ropes and as the rock was lifted out to be thrown away, it seemed unusually heavy for its size. One of the men broke a piece off and put it in his pocket, throwing the big chunk away. George Brown's advice was requested as to the kind of rock it was and he found that it ran very high in gold. George, like so many others, had been too busy to go up there and look for the chunk that had been thrown away and also its source, but after about ten years since the telling of the story, he hasn't been able to check on it.

(199)

About the gold taken by Stage Coach Robbers in 1856, Rattlesnake Dick Barter, Tom Bell, Dutch Kate, Black Bart, Murietta, Vasquez, are only a few of those participating in stagecoach stickups. However, they garnered considerable wealth in gold dust, coins and bullion; and few bandits lasted long. In many cases, their ill-gotten gains were hastily buried. Starting soon after the gold rush in California, a few samples of the amounts involved will show the possibilities of treasure hunting. Sums like the following were reported lost: $7,500 being the first of consequence in 1852; $15,000 $40,000, $45,000, $100,000, $150,000 and $200,000. A hoard of $40,000 is reported to be buried on Trinity Mountain, California and one of $150,000 at Jackson Hole, Wyoming.

(200) "Locating Treasure"

Many devices are used such as the cartucho — a metal cartridge shell filled with opium, poison, black rock and other elements which hangs from the bearer's forehead and oscillates in the direction of the metal; the horquetta — a forked scapula from the sheep, the upright branch fitted into a glass knob, on top of which is a threaded hole, an assortment of hollow screws fit into these threads, one is filled with gold dust, another with copper, the choice of metal for each operation corresponding to the metal sought. To work the horquetta, the lower prongs are grasped in the two hands, thumbs out and palms turned up. The horqueta should twist and strain in the direction of the sought-for mineral as if drawn by a powerful magnet, and then, when over it, should point straight down in the manner of the water switch that turns bands toward water. Many magnetic devices are in use and the "Geiger Counter" is very valuable in locating uranium. It is so attuned that it registers cosmic rays at points far underground and sets up a lively ticking when atomic particles from radio active material reach the counter.

(201) "Comanche Chief Yellow Wolf and Great Silver Lode"

The Comanches were a fierce and energetic people. They held contempt and even hatred for the Spaniards. They were strong enough to evict the Apaches from the territory of the Llano-San Saba and fought against the settlement by Anglo-Americans. During the period 1851-1852, Chief Yellow Wolf was at peace with the white man. His tribe at the time made the North Fork of the San Gabriel River in Williamson County their headquarters, the camp being about two miles above the Mather place. The Chief took almost pure silver ore to Mather to be hammered into ornaments. The deposit was three suns west of the Indian Camp, under a bluff near the junction of two streams. Old Fort San Saba would be about three days' travel to the west.

A former captive of the Comanches located the silver lode thusly: Starting at the Old Presidio, the Indians crossed the San Saba River and followed up Los Moros Creek. A line projected southward from the old Presidio up Los Moros Creek would nearly coincide with the line of silver-bearing sandstone that Captain George Keith Gordon traced to the Lechuza Ranch. A silver ledge on the Frio River is another of the sources for material for bullets and ornaments. It is on the south bank of one of three arroyos that run into the Frio River close together.

A boy captive of the Comanches, who lived in Burnet County tells of silver nuggets being picked up in a cave in the form of pebbles. This cave should be about 50 miles from Burnet. Then the silver may some day be located by finding a thick mottle of persimmons which contains a broad rock slab. A slanting hole underneath, and within the hole a myriad of glistening silver stalactites hanging from the roof. Beasley's Cavern may lead to the silver lode. Beasley, with other settlers about Lampasas, took up the trail of a raiding band of Comanches, which took them across the Colorado and North toward the San Saba. Near camp one morning, Beasley discovered the cavern, which faced east with the rising sun shining directly into it; the hole sloped down at a steep angle, then leveled off into a horizontal tunnel; the tunnel walls were lined, plated and cased with almost pure silver. This cavern is near the mouth of the San Saba River. The Comanches knew of these silver deposits and the legends tell us that when old Chief Yellow Wolf died, his treasure was buried with him. Near Monahans, under the drifting sand dunes, his grave will some day be restored to sight by the change of wind and the three mule loads of gold and silver will be brought to light.

(202) "Buried Plunder" 1884 — Arizona

Robberies in old Mexico netted about $3,-000,000 consisting of a cigar box filled with diamonds worth about $1,000,000; thirty-nine bars of gold worth $600,000; and $90,000 in silver coins; gold statuary figures and stacks of gold coins.

Curly Bill Brocius, a robber baron who operated near Tombstone, Arizona, was aware of mule teams bringing smuggled riches into Arizona from Old Mexico. Several of these pack teams were attacked by Brocius and his gang in the San Simon Valley, murdering the smugglers. Jim Hughes, a member of the gang, could speak Spanish, so was sent to Sonora, where, posing as a Spaniard, learned about the departure of a pack train which would pass through Skeleton Canyon and San Simon Valley. Reporting back, Hughes found his chief away, so he decided to engineer the job himself. The smuggler's pack train consisted of fifteen men and thirty mules. They entered Skeleton Canyon and posted guards when they halted near Devil's Kitchen for lunch and a siesta. Rifle fire from the canyon walls cut the guards down quickly, the others regained their mounts, but four were shot out of their saddles. The remaining Mexicans fled leaving the treasures behind. The loaded mules scattered, but were shot down one by one until all the richly laden packs were recovered. The loot was assembled in one pile so high they didn't have horses enough to carry it, so decided on a temporary place of burial.

In recent years, coins scattered by the frightened mules have been found. The job of removing the treasure to its more secure resting place was assigned to two trusted helpers, Hunt and Grounds. Soon after this was accomplished, Hunt was wounded and Grounds killed in a gun fight in Tombstone. Hunt was now the only one knowing of the new burial place. The wound that Zwing Hunt had received proved his undoing, but before he died, he told an uncle of the buried treasure. Prior to the raid in which Hughes participated, Grounds and Hunt had robbed smugglers.

This loot, together with the last haul was removed by them with the aid of a Mexican teamster who was killed to seal his lips. Hunt's map to his uncle set down Davis Mountain, Silver Spring, Gum Spring, a rock with two crosses and a spot near the San Simon. Nearby, a waterfall in the Canyon affords a good place to bathe. Hunt's Mountain is possibly only a mound where one of their pals named Davis was buried with $500 in his pockets. Hunt's uncle, Ground's brother, the Chief of Police of Tompstone, Porter McDonald and many others have searched in vain for the treasure.

(203) "Lost Peg Leg Mine"
1829 — California

Awaiting the finder, some black knobs of the Colorado Desert, constantly burned by the sun, and washed with flash cloudbursts, have on their surface quantities of gold nuggets so altered as to make their identity hard to recognize.

Thomas L. Smith (Pegleg) Smith ran away from his Kentucky home to become a trapper. The year of 1823 found him with the Alexander Legrand party at Santa Fe, New Mexico. Then in 1824, he joined a trapping party in Utah to work a tributary of the Grand River, which was afterwards named after him. Smith, LeDuke and Sublette worked together in Apache County on the Gila River. Sublette was wounded by an arrow here. Smith received the wound which caused an amputation — a rough job done on the spot — from an arrow, while on the Platte River. In the spring of 1829, Smith and Ewing Young, heading a party of trappers, camped on the north bank of the Colorado River, a few miles east of its junction with the Virgin River. Dutch George, one of the trappers, showed Pegleg a couple of yellow nuggets which were taken for copper. It could be used for bullets, so some of the men went downstream to a point about two miles above the Virgin River. The Dutchman said the bottom of the dry creek bed was covered with the stuff, but he became confused when draws of similar type were crossed and he was unable to take the party to the right wash. Being just copper, they didn't search very long anyway.

When it came time to get the furs to market, Pegleg and Maurice LeDuke were chosen for the trip. Their route is supposed to be down the Colorado as far as the junction with the Gila River near Yuma. From this point, they sighted a range of mountains to the northwest and after three days with water about gone and part of the fur pack buried to lighten travel, they came to a mountain that had vegetation but no water. From this point, three small buttes could be seen — two close together and the other by itself. Pegleg decided he would try for water between the two, but found none and climbed to the top of one to view the surrounding country. He was sitting on stones which bothered him, so he threw several of the small ones away. He then noticed that one which he was about to heave seemed unusually heavy. He struck it against a large stone, breaking off the dark coating, and discovered it looked like the stuff Dutch George had found. Smith put several pieces in his pocket. The fourth morning out, they saw a mountain to the northwest which they had used as the original site and found a spring of cool water. They remained here a day, naming it Smith Mountain (not the Smith Mountain changed to Palamar), then moved on over the mountains to the Coast and to Los Angeles where

they were told the stuff found by Pegleg was gold. Smith was ordered to leave town after beating one of the citizens over the head with his pegleg during a drunken fight. He joined Bridger and Sublette in search for furs again, this time on the Grand and Bear Rivers in Utah, about 1849. Twenty years had elapsed, but he was talked into making a search for his gold on visiting Los Angeles again. A party of three left for Warner's Ranch as a jumping-off place. A drinking bout that lasted four days and trouble with the Indians who insisted on driving off their livestock, confused the party, so that nothing was gained. Pegleg later resorted to asking for grubstakes. He finally interested a large party and they set out to look for the gold on the Colorado near the Virgin River. Pegleg claimed they couldn't "take it", never reached the place and returned to Los Angeles about two months later.

About 1855, a freighter named Brady had passed through Box Canyon and entered the flat country west of Sentenac Canyon, where he came upon Pegleg clinging to the pommel of his saddle, almost unconscious, and letting his horse drag him along. He talked of a black butte strewn with gold, and showed a couple of nuggets as large as marbles. He said he had been living with Indians at Yuma in hopes of gaining information of gold from them which would help him locate his black butte. He married a squaw, but to no avail, then tried again but failed. He told them at Warners,

Stage Station, Warners Ranch, in the heart of the lost mines area of San Diego County.
— California Department of Public Works

where he recovered from his experience, that his former partner, Maurice Le Duke had been killed on the Gila River.

Julio Ortega, who has lived at Warners Ranch for 60 years, said his father, Antonio Ortega, received directions from Pegleg as to the route of himself and Le Duke which was simply that they kept coming west after they left the river, heading for a low-lying range of hills that didn't have water, from where they saw the black hill and found the gold; and that the next day they turned more north and found a spring at the base of mountain. Gossip and clues place the "Lost Pegleg" in the Cocopahs, south of Signal Hill, just over the United States-Mexico border; in the Fish

Mountains (named Smith Mountain by Brandt) near Carrizo; in the Little San Bernardinos, near Twenty-nine Palms; and in the Chocolate Mountains.

(204) "Indian Nuggets" 1870 — California

An Indian, fond of gambling, runs into hard luck at the rendezvous for bad men at Puerta de la Cruz on the old Butterfield route, north of Warner's Ranch. To recuperate his losses, the Indian appeared later with a buckskin bag of nuggets. When asked where it came from, he replied "Maybe from Bucksnortes", (some small peaks east of Oak Grove near Collin's Valley. Search did not reveal gold, but its presence led to other clues). The Indian's squaw became seriously ill and was cared for by Mrs. Wilson, the storekeeper's wife at Warner's Ranch. After recovering, to show her gratitude, the squaw promised to take Mrs. Wilson to the place where her buck found the nuggets. Halfway down the San Felipe, Mrs. Wilson played out, so the squaw told her to rest in the shade while she went on to the mine. She went east into the grapevine Hills and returned a few hours later with a handful of nuggets for Mrs. Wilson.

(205) "Charles Knowles and the Dutchman" — 1872, California

Others have found black stones on the desert that when divested of their desert varnish revealed gold. Charles Knowles, an oldtimer of the Idaho Panhandle, was told of the Pegleg Gold and decided to make a try for it. He moved his family to Mesa Grande and ranged from there. While camped at Carrizo Creek he viewed the strange sight of seeing a dirty, unkept, bearded man wading down the creek, staggering along like a drunkard but looking straight ahead with unseeing eyes. Knowles, realizing he was delirious, carried him back to his camp and cared for him until he recovered. The man had walked from Yuma and was heading for Los Angeles. While talking of his experience, he reached in his pocket and produced four small black stones. Two pieces proved to be barren quartz, but the other two showed gold nuggets attached to the quartz and iron. The man couldn't remember where he picked them up and refused to stay with Knowles to do some prospecting. Knowles spent considerable time prospecting the Carrizo-Fish Mountain area, but could not find the black nuggets.

(206) "Peg Leg's Ring" 1870 — California

"Slue-foot" Brandt took $25,000 out of a secret canyon in the Fish Mountain area. Legends carry Pegleg Smith's activities beyond his time, so perhaps "Slue-foot" is another Pegleg. Henry Miller of Live Oak Springs, tells of his father freighting between San Diego and Yuma for several years. On three occasions, he was stopped by a peglegged prospector who asked for a ride down the mountain from Jacumba Valley to the desert. The prospector carried only enough food to last five or six days. On the last trip down, he handed Mr. Miller two large gold nuggets and thanked him for his kindness, saying that he could get plenty more. Miller had one of the nuggets made into a ring. Miller's freight route was from San Diego southeast across Otay Mesa into Tijuana Valley, east through Chollas Valley, Valle Redondo, La Puenta, San Jose, Valentin and Canyon Verde to Jacumba. From Jacumba northeast and north past Table Mountain to the desert mountain edge overlooking Mountain Springs, at the foot of Devil's Slide. "Slue-foot" would get off at Devil's Canyon, where he would sit down by the roadside until Miller was out of sight. Henri Brandt died in San Bernardino in 1914, but before this described his mine as "Pegleg's Ring."

"I can go west from my mine to a hot spring above which is a double-decked cave. I can go south from the mine to a painted canyon. To the north I can come out of the hills through Split Canyon. From Split Canyon, I walked to Kane Springs. When I went to my mine from the brush shack east of Carrizo, I passed a cave filled with Indian ollas. From the shack to the mine took four hours. This route is marked by two human skulls on ocotillo stalks about a mile apart. There is seepage just below the mine."

(207) "Desert Gold" 1902 — Lower California

Out of Agua Jito (Little Water) between Rosario and San Fernando in Lower California, is the J. U. Smale Turquoise Mine. An employee, Antonio Joseph, is shown the hiding place of desert gold brought from the north by Indians. While prospecting in the Sierra San Pedro Martir, he left the mission and headed toward Santa Catalina. On the third day out, while crossing a small mesa, someone called to him. It was an injured Indian. Antonio helped him to his Indian village and in thanks the Indian offered to show Antonio something no other white man had ever seen. They went to a narrow box canyon with walls a hundred feet high. The floor was barren, but at the end of the enclosure a chiseled rock pedestal supported a covered stone image of an Indian which faced the east and when the sun shines upon it in the morning the cast shadow points to the Chimenea de Ceniza (Chimney of ashes) which contains the stored gold brought from the north.

Antonio had agreed to be blindfolded on the approach and departure from the box canyon, but he received instructions on how to get to the source of the gold. "From here you must cross the sands to the mountains where the Indians hunt sheep (the Cocopahs) follow the big range north until it ends in a big mountain (Signal Hill); cross the desert for a day to another mountain and in a canyon facing east you will find gold." About six months after Antonio left his employer, J. U. Smale, some surveyors, while working around Split Mountain Canyon above Fish Mountains, found the remains of two mules, which had been tied to stakes, and some large water cans, but no trace of the owner.

(208) "Lost Borego Mine" 1890 — California

On the western side of Borego Valley is a remarkable outcrop of gold ore. First it was necessary to find a large flat-surfaced rock imbedded in the side hill with spearhead carved in the rock with a polepick.

A Los Angeles school teacher, daughter of a German miner, enlisted the aid of Charles Knowles in the search of a mine left her by her father. He had left her $4200 in gold, several ore specimens and a map. After ten days' search, the lady returned to Los Angeles, but Knowles continued the search later and found only the remains of an old arrastre on Coyote Creek where the old German had probably crushed his ore.

The Borrego Valley, first traversed by Anza and his party in 1774, is the site of the vast gold ore deposit sought by a Los Angeles school teacher whose father, a German miner, left her $4,200 in gold, ore specimens and a map.

— California Department of Public Works

(209) "Peg Leg Smith the Second" 1869 — California

Find a mine so rich that every street in San Bernardino could be paved with gold, the one pointed out to Smith by an Indian squaw who had been injured in a fall and was dying of thirst and hunger. He took her back to her people and was rewarded with a handful of nuggets and directions as to finding the mine. Smith was a hard drinker and once when on a rip-snortin' bender he found it too hot in his cabin, so climbed to the roof and went to sleep. When discovered later by one of his friends, the cry "Fire" rang out, and Smith, foggy and bewildered, staggered about, then fell off the roof. Whether this story accounts for his dying isn't clear. However, later, after a strong drinking bout, he became ill with pneumonia and thinking he was about to die, told his nurse in San Bernardino his trail route, as follows: "Down a valley to a pass leading east, turn north to a spring, then north to a ridge of small hills, one of which looks like a hog's back, and you have it." The pass could be San Gorgonio Pass, the spring Cottonwood, and the pass between the Little San Bernardinos and the Orocopia Mountains. Recently, in January, 1946, the writer talked with a mining man operating in the vicinity of White Tank—a point on one of Smith's alternate routes just south of Twenty Nine Palms, who said a strike had been made in the Pinto Mountains. An old timer, over seventy, working along for years in an abandoned mine, one tunnel showing evidence of being at least fifty years old, had at last decided to quit as the digging was producing less every day. However, he would have one last try at it with his jack hammer and this time he would go up toward the top of the mountain instead of down. To do a good job his drill holes were long and numerous enough to take a good charge. After the explosion, upon inspecting the material, he was amazed and delighted to find a large amount of quartz studded with stringers of free gold.

The word has gone out that the "Mother Lode" of the Pintos has been found.

Note: —According to an article in the Desert Magazine (1950) a systematic search is going on in the Pintos for the Peg Leg "black gold." It seems that a mule skinner used "black gold" to throw at the mules he controlled with a jerk line. His granddaughter is making the search.

(210) "The Lost Candelaria Mine" 1869 — Lower California

About 65 or 70 miles southeast of Ensenada lies a mineral district which has produced several millions in gold. It is also the locale for a lost gold mine—one that didn't produce "fabulous wealth" but was a starter.

Ensenada is about 70 miles south of Tijuana on the Bahia de Todos Santos—(Shoals of All Saints). Early navigators, pirates, adventurers of all sorts and mining people knew the port.

A young San Franciscan bank clerk of French ancestry found it advisable to leave his job, and headed for Ensenada, having heard of the activity in gold. He hired two Indian guides from La Hurta and worked for six months around Las Cruces Canyon and Real del Castillo. Then he dismissed his men and went alone into the Candelaria Canyon. When he showed up at Ensenada, he had several bags of high grade ore, each bag being marked "Candelaria."

He took passage for home but never reached there although the ore was delivered to his family, who received $12,000 for it.

Many have searched for the Frenchman's Mine but failed to find it. The Viznaga district — Santa Cara Placers—produced ore assaying $10,000 to the ton. The Alamo Creek diggings, southeast of Santa Clara, produced $2,500,000 in gold. A miner drove a tunnel into the hill north of Alamo Creek that netted $1,000 per foot of tunnel. This tunnel is undiscovered.

(211) "The Lost Bull Ring Mine" 1836 — California

Rich gold ore awaits the finder in a shaft covered with rough timbers of pine logs that look like driftwood, one of which contains a driven iron staple to hold a brass ring.

Go back to the Black Hills near Mule Springs Canyon, about fifteen miles west of the Colorado River. In a foothill canyon on the north slope near a thicket of stiff stubby bushes you should find the shaft.

Very little ore was taken out by the Frenchman who discovered it, but one of his countrymen who had been told of the find took out several loads that were transplanted by boat down the Colorado.

A man and his wife, with some children, travelled the old desert road between San Bernardino and Ehrenberg about the year 1900. They camped late one afternoon near the end of their journey and about the time supper was ready one of the children ran into camp saying, "I have found a well with a ring in it." The father paid no attention to the child, but later hearing of the "Lost Bull Ring Mine," he tried to locate it but failed. They might have camped at Mule Springs or at Blythe Junction or Rice.

(212) "The Aztec Montezuma" 1518 — Mexico

Many loads of treasure—gold and jewels in the form of an enormous cache, is supposed to lie buried in Northern Sonora, east of Tinajas Altas near the United States boundary line and Ajo Mountains.

This great wealth, representing half that possessed by Montezuma, came from the Sacred Valley of Anahuac. The Spaniards had landed at Vera Cruz, showed their avaricious nature when gifts were sent them and caused Montezuma to fear for the safety of his treasure.

(213) "Guadalupe de Taiopa" 1842 to 1873 — Mexico

Nuggets as large as pears by the dozens—so many that in a very short time one could pick up a large bag full. Taiopa ranks with Tarasca as a Great Mine.

To find this mine look for the ruins of Guadalupe de. Taiopa. East of Nacori are small ruined settlements in the heart of the Sierra Madres. John Calhoun Smith, a nephew of Pegleg Smith the trapper, thinks the mine is at the headwaters of the river Yaqui near the ruins of a church in the mountains of Chihuahua,, and that the Veta Madre (Mother Lode) is near Sahuaripa.

About 1842 an expedition hunting Apaches came upon a lake which is on the westerly side of the River Yaqui. There existed the remains of the Sierra Madre Mountains and headwaters of an old furnace on a small stream close to the lake, dross of silver and copper, old mine shafts, and in a hollow a member of the party found a piece of gold weighing ten ounces. Around the lake Indian tobacco grew in quantities. This is between Sonora and Chihuahua States. An old Mexican woman of Ures knew of the place.

(214) "Treasure of Don Juan the Mad Conquistador" 1810 — Mexico, California

Don Juan's personal treasure and "muy rico oro" from the Taiopa lies buried in an unknown arroyo between Signal Hill and Carrizo.

Forced out of Mexico during an uprising, Don Juan set out for Monterey, California. He had sixty horses — about half of them loaded—and took along twenty-four Indian servants. His route from Guadalupe was westward to Minaca, then north into the Sierras to the headwaters of a stream near Mulatos; thence into the hills of Sahuaripa and then to Oposura on the left branch of the Yaqui River and to Nacori.

Between Sahuaripa and Oposura the extra animals were loaded with rich ore. An old Mexican claimed that when Don Juan reached the Casa de Peralta in Oposura, one of the servants boasted of "muy rico oro." The party reached the Colorado near its junction with the Gila and made the dangerous journey to the first water of Carrizo.

Trouble overtook them near Signal Hill in the form of a large band of Indians bent on killing them. A running fight took place and the treasure was all piled together in a small arroyo where the last stand took place—somewhere between Carrizo and Signal Hill. A Spanish breastplate of the period 1770-1780 was found near Signal Hill in 1937.

(215) "The Lost Mine of El Moranjal" 1800 — Lower California

A large pile of enormously rich ore lies in a canyon below La Mesa Encantada, "The Enchanted Tableland" an inaccessible place, with steep, high walls. Early Spaniards found traces of an ancient civilization, of buildings and irrigation ditches and stumps of what appeared to be orange trees. These Spaniards mined in a big canyon that headed near the mesa. The ore was thrown into a great pile against the day when a friend would come up the gulf with a boat to transplant the gold to the south. The boat didn't arrive, supplies ran short, then the Spaniards found themselves on the wrong end of a revolutionary movement and had to quit the country.

Ore brought into Tiajuana, Lower California, for assay, ran $25,000 to the ton. According to the half-breed who found the ore, it came from a deep canyon, was already mined and in a pile as big as a house. Guaymas was the nearest settlement, the largest canyon formed from many smaller ones leads to the gulf at a point about 100 miles below the head of the gulf.

(216) "Jamatai" Gold - Jacumba 1682 — Lower California

Many legends point to the fact that white men lived in the mountains south of Jacumba and disappeared long before the country was colonized. They tell of the Indian chief who ruled the valley and had a place where he kept an olla full of gold ore so rich that he had a necklace made from pieces of pure gold as large as mesquite beans.

Another Indian chief used only gold for his arrowheads.

Then there is the rich treasure of placer gold that was dug by members of the wrecked Isabel la Catolica. Their ship was wrecked in the year 1682 on the Lower California Coast near Bahia de Todas Santos (Bay of All Saints) at Ensenada. The crew having provisions for only five days decided to go into the mountains and find game. After traveling north-easterly, they came into a fine wooded country with game and water.

One of the men discovered gold in the stream bed near camp and they began to wash gold; as the season was wet, and kept at it for eight months. The men were attacked by a large band of Indians but before all were killed one man left a manuscript with directions for finding the gold.

In 1873, Pedro Padrillo found the manuscript in an iron box three feet below the surface, as directed by an inscription on a cross found at the surface. There is no record of Pedro finding the treasure.

Three Mexicans found treasure in the San Miguel Mountains. Their route was from Ensenada easterly, 30 leagues to a spot marked "Aqua Fosforescente" thence south ending in a conical peak covered with huge stones from which vantage point they could see another mountain far to the south. When its base was reached they located an old trail which led them up the mountain to an old tunnel, the mouth of which was filled with stones and old timbers.

Hand picked samples of gold ore from the ancient dump in front of the tunnel gave them a sufficient amount to load eight animals. About a day's time from the tunnel they camped in a dry wash about a mile above some sulphur springs. A quarrel among themselves left but one survivor but the Indians drove his animals away during the night, leaving him afoot. He buried his two friends and the treasure and returned to San Diego.

In 1873, placer gold was found 80 miles from San Diego at Japa. Leave the Yuma, Tecote road at St. Valentine and go to Las Juntas, Agua Chisera, Templados and Japa.

(217) "The Treasure of Santa Isabel" 1782 - 1860 — California

Gold, and Indian amulets lie buried on the eastern slope of the Mesa Grande Mountains.

The valley of Santa Isabel lies about fifteen miles south of Warner's Hot Springs and six miles west of Julian. The first white men to enter the valley were those under Don Pedro Fages and Father Juan Mariner. The location for the mission was decided in 1795, but wasn't built until 1822, by Father Fernando Martin. Two fine old bells, dated 1723 and 1767, were stolen, supposedly by older Indians displeased with the actions of the younger ones. They may be buried with the gold.

Three missionaries joined a caravan of homeseekers headed for California about the time of the gold rush. They were heavy drinkers, and becoming troublesome, a "drumhead court" ordered them to leave the caravan. Coleman Creek was the source of their placer gold. Then they entered the valley of Santa Isabel from the south, they kept to the western side hoping to avoid attention, but were discovered by Indians, one of whom was set to spy on their movements. They were watched when during the division of the gold a quarrel developed in which two of the men were killed and the third was killed by the Indian.

The gold, consisting of nuggets in a bag of deerskin as big as a man's head, was taken to the chief. That night he carried it across the valley. Indians intimate that the gold, the bells and Indian treasures were buried at a spot that could be seen from the front door of the church and guarded by "eyes." The Indian shamans being able to preserve human or animal eyes, breathed the spirit of guardianship into a necklace of them and placed them over the cache as protection.

(218) "The Black Crow Mine" California

Enormously rich ore — a badly shattered quartz in which gold shows in every seam — lies within a two-mile radius of a point on the north wall of Blair Valley.

This might be another Lost Dutchman's Mine, as the man who found the ore was called the Dutchman, being German and with the name of Jasper Dietz.

Dietz set out from Yuma to do some prospecting for the Lost Pegleg Mine. He had heard many stories of this lost mine while employed in a Yuma saloon and felt sure he could find it. He was outfitted for the trip, but while trying to take a short cut his mule slid into a canyon, dying from the injuries received, and lost the precious water to the dry sands. Old Man Blanchard of Sentenac Flats rescued Dietz from the threatening position he found himself in, nursed him back to health, and they made a deal as to further prospecting. A brush hut was constructed over in Blair Valley which would be nearer to the area to be worked and Blanchard was to furnish grub and water. Dietz was to report at Blanchards every two or three days. The old man came in several times and on one occasion mentioned having found a stringer that looked good, so he dug on it. He hadn't gone but a few feet when he came upon a flat square stone which he lifted and as he did so a black crow flew out. Dietz said "that Crow was the spirit of Pegleg Smith and will lead me to a great discovery, as I want the money to take care of all the orphans of Belgium." After telling Blanchard of the stringer of gold, Dietz left camp again but when five days elapsed and he didn't show up, Blanchard and a couple of visitors set out to look for him. They found the old man feverish and very sick lying on the floor of his brush shack but he asked Blanchard to look at the specimens he had brought in. Many worthless rocks slipped from one hand to the next as they were examined, but finally a large piece showed the free gold in its many seams. When Dietz was asked where it came from he could only say, "to the east", and that it was covered up by him, but that there were tons of it. Dietz died in San Diego.

(219) "The Lost Mission of Santa Isabel" 1800 — Lower California

When the "Politicos" and the Mission Fathers disagreed and the padres knew they must leave the country, they decided to build Santa Isabel, which was to hold the treasure of all the missions. After this was placed in the walls, trees and cactus were planted in the path so that all signs of the trail leading to Santa Isabel were destroyed. The mission is at the base of an impassable cliff 7000 feet high, but there was an entrance from the desert which had been blocked by landslide. A Yaqui Indian, while looking for the Mission Santa Maria, was given the wrong direction by another Indian. His route was northeast out of the San Gustin to the small valley where three trails were before him. He took one leading to a higher level. It ended in a sheer precipice. Below him, far down, lay a narrow canyon, and in a hidden valley at its mouth stood a fine mission. Chaparral grew near the door, bells hung on a cross bar over the door and to the rear were some natural water tanks and three large palm trees, arrowweed and tobacco plants. From a high point near the mission the Indian could see the peak of San Juan de Dios and knew he was far north of the trail he wanted.

The Mission Santa Isabel is in a district rich in gold and silver. The Yaqui Indian, according to legend, received the curse of the Mission when he revealed its hiding place and was struck dead.

Uranium!

THE MOST FABULOUS MINING BOOM SINCE THE GOLD RUSH DAYS

PROSPECTOR

MINERS' REWARD!!

The U. S. Atomic Energy Commission will give a reward of many thousands of dollars for new discoveries of high per cent uranium.

CONTENTS

URANIUM . . . INTRODUCTION

An attempt has been made to condense and correlate the maze of existing material. State and Government circulars have been utilized. The brochures furnished by the U. S. Atomic Energy Commission have been scanned and condensed. Gems of vision delivered by men of industry have been utilized.

The action relating to uranium is very fast and ever-changing. The incentive programs of the Atomic Energy Commission both in time and price schedules allay any doubt as to the permanence of the new industry. In the price schedule for instance—0.10% grade ore is worth $7.00 per ton plus a haulage allowance; 1% grade ore is worth $164.50 while 10% is worth $1694.50. As for the time schedule the producers will have until 1957 for one phase of the program and until 1962 for another.

The search and discovery of radio active minerals will result in a wonderful atomic age. It may be called the "Era of Atomic Creation." Its horizon will be limited only by clear vision and capability. Potential energy reserves of the atom era are 20 times existing fuels such as coal, gas and oil. 1 pound of uranium 235 has the energy value of 2½ million pounds of coal. The potential horsepower as a factor of work may increase one hundred times. Streets will be lighted and the homes and buildings heated by the atom. Trains, airplanes, ships and submarines will be powered with fissionable material. The waste matter of cities will be disposed of. Sea water will be made available for agriculture and industry. The limits of economic transmission and transportation will be a thing of the past. Substitution will take place or effect every familiar object we now deal with because of the portability of fissionable fuels. The time when atomic energy will become competitive can now be predicted. The magnitude of the resulting wealth produced will be fantastic.

URANIUM . . . MODE OF OCCURRENCE

Uranium occurs principally as the oxide. The oxide is found associated with pitchblende, as the sesquioxide in inaconite or uranochre. It occurs as phosphate in uranite or uranium mica, as phosphate of uranium and lime in autunite, as phosphate of uranium and copper in trobanite, as carbonate of uranium and lime in liebigite, and in trogenite, uranasphaerite and samarskite.

Uranium ore deposits occur in all major types of rocks — igneous, sedimentary and metamorphic. In igneous rocks, uranium occurs in hydrothermal veins, in disseminated bodies and in pegmatites. In sedimentary rocks uranium occurs in carnotite deposits, in the phosphorites, ashaltites and limestones.

Of the commercial varities uraninite pitchblende and carnotite are the mose important.

URANINITE: Uraninite may contain radium, lead, rare-earths, thorium and helium. It is usually found massive and botryoidal and also in grains. Its luster is glassy-dull, its color brown to black. It has been found in hydrothermal veins with silver, lead and copper.

PITCHBLENDE: Pitchblends contains a low mixture of thorium or rare earths in the uranium oxides that occur as vein material associated with sulphides. It has the same general composition as uraninite, a pitch-like luster, a fine dark greasy like powder that will smudge the finger if it is in contact, it occurs in sulphides and quartz, and it is associated with other minerals especially copper sulphides and hydrocarbons.

CARNOTITE: Carnotite is yellow, hydrous, potassium uranium vanadate with the orthorombic crystal system. It occurs as a yellow powdery incrustation or as a cementing agent in sandstone. It is associated with petrified wood, fossils, bones, malachite, azurite, biotite, magnetite and secondary uvanite and tyuyamunite.

Uranium is distributed in many types of rocks. Study shows that pegmatites contain more radioactive minerals than other igneous rocks, however the great distribution dilutes the values.

Those rocks high in potash show more radioactivity than the basic rocks gabbro and basalt. Accessory and secondary minerals may account for much of the radioactivity in granite, quartz monzonite or granodiorite. The phosphorites, limestones, black shales and asphaltites may show uranium. The uranium-rich carnotite ore bodies are largely throughout the Salt Wash member of the Morrison and Shinarump formations. These bodies range from small masses a few feet in dimension to large tabular bodies several hundred feet. The beds are largely horizontal but may be tilted. In exploration of these ore bodies for structural control careful mapping and projection is in order. The copper-uranium minerals are largely in the conglomerate or sandstone of the Moenkopi, Shinarump, or Chinle formations. For the most part uraniferous-asphalt occurs in the Shinarump formation.

KNOWN URANIUM FORMATIONS

CENOZOIC ERA: Eocene Wasatch—3000 feet; Eocene Mesa Verde —1000 feet.

MESOZOIC ERA: Upper Cretaceous—Dakota 50 to 200 feet, gray; Manos Shale—2000 to 5000 feet; Burro Canyon—50 to 250 feet, green, maroon. Upper Jurasic—Morrison—300 to 500 feet, vari-colored; Entrata—50 to 1000 feet, gray or red; Summerville—50 to 400 feet; Curtis, 0 to 250 feet; Salt Wash—200 to 400 feet, light to red; Carmel—0 to 600 feet, orange or red; Navajo 0 to 2000 feet, light color; Kayenta—0 to 300 feet, red; Wingate—0 to 400 feet, red; Chinle—100 to 1000 feet, red, lavender, green, gray and pink with fossil wood; Todilto formation carries pitchblende, carnotite, tyuyamunite and uranophane. Triassic—Shinarlump Conglomerate—0 to 50 feet, white to dark with petrified wood; Moenkopi—0 to 1000 feet, chocolate color to red.

PALEOZOIC ERA: Upper carboniferous — Pennsylvania Phosphate rock; Permian—Cutler and Rico formation—0 to 6000 feet.

PRE-CAMBIAN ERA: Sediments — varied thickness — copper and uranium.

COLORADO PLATEAU HOST ROCKS

Especially favorable host rocks and the greatest source of uranium on the Colorado Plateau are: Mesozoic Era Triasic Moenkopi, Shinarump and Chinle; and the Mesozoic Era — Jurassic Morrison with the Salt Wash Member. Carnotite is found in all of these. The Temple Mountain and San Rafael Swell districts produce the dark asphaltic uranium ores from the Mesozoic Era—upper Triassic Period in Shinarump Conglomerate.

The Colorado Plateau consists of a thick series of sedimentary beds of many colors and types. These beds are horizontal except where pierced by igneous intrusions and subjected to forces resulting in anticlines, folds and faults. There are fourteen known formations which contain uranium deposits. Most of the uranium deposits are in sandstones, limestones, and shales parallel to the bedding. Some fissure veins containing uranium are known.

The deposits extend from near Myton, Utah, and Skull Creek, Colorado, to Grant, New Mexico, and to Holbrook, in Arizona; and from Meeker and Rifle in Colorado, westward to Leeds in southwestern Utah.

URANIUM BEARING MINERALS

Uranium combines with other substances to form a mineral. In the 100 or more known uranium bearing minerals uranium may predominate or it may be a very small part of the whole.

A knowledge of the minerals associated with uranium may lead to its discovery in many places. A list of the most important uranium bearing minerals is given with some of their associated minerals.

Autunite—calcite, phosphate, vanadium, chromium, mica, chalcopyrite, azurite, asphaltum, malachite, molybenite.

Betafite—cblumbite.

Brannerite—titanium.

Carnotite—potash calcite, quartz, fossil-wood and bone, chromium, vanadium copper, selenium, iron, silver, lead.

Copper—uranium type—cobalt, nickle, arsenic, iron, lead, phosphorous.

Coffinite—arsenic, galena, chalcopyrite, chalcocite and eucairite.

Euxenite—titanium, columbite.

Fergusonite—lead.

Lumsdenite—vanadium, uranite, and copper-silver-selenide.

Monazite—rare earths and phosphorous.

Pitchblende—rare earths, quartz, pyrite, rhodochrosite, calcite, galena.

Roscoelite—chromite, lead, vanadium.

Samarskite—rare earth colmubium, iron, calcite.

Thorianite—thorium.

Thucholite—carbon.

Torbernite—copper and phosphate.

Tyuyamunite—vanadium, calcite.

Uraninite and Uraninite Hydrocarbons—chrystaline pitchblende, black vanadium, cobalt, pyrite, arsenic, and selenium.

Uranophane—calcite and silica.

Uraniferous Asphalt—selenium, copper, arsenic, cobalt, iron, vanadium, metallic sulfides, corvusite, pyrite, carnotite, realgar, zunnerite and crythrite.

Uraniferous Hydrocarbon of the Dolores formation has the following associated minerals—barite, calcite, chalcopyrite, copper carbonates, erythrite, pyrite, spalerite, chromium mica and tetrahedrite.

ORIGIN OF URANIUM ORE

The study of the origin of uranium ores continues with changing concepts. It is now agreed that the deposits of black ores are extensions of the oxidized and hydrated carnotite type minerals which are mined at the surface. It is now believed that the uranium mineralization of the Colorado Plateau and the Colorado Front Range is of hydrothermal origin. Areas containing selenium bearing plants may be mapped as indicators and a guide to uranium bearing ores. Loco weed, other plants and juniper may contain selenium.

To account for the presence of uranium in petrified wood the hypothesis is proposed that assumes older uranium and vanadium veins in the drainage basins that supplied the La Plata and McElmo sediments of the Jurassic Period—southwest Colorado, furnished the mineral. These minerals were disolved by sulphuric acid generated by pyrite and carried to the sea, where they were precipitated by decaying reeds and trees.

REGULATIONS GOVERNING URANIUM AND OTHER MINERAL LOCATIONS

Locating a claim.

(1) DISCOVERY. A mineralization seam, vein, fissure, fault, or disseminated mineral which may lead to an ore body constitutes mineral in place.

(2) LOCATION NOTICE. Place same at point of discovery. Name of claim. Name of locator. Date of the location. Length and width—600 x 1500 feet. Course of the claim. Reference to some natural object.

The location must be marked on the ground within 90 days for a lode claim in Arizona, California, Colorado, New Mexico, Wyoming and Utah and within 20 days in Nevada.

A stone monument 3 feet high or a stake at least 4 feet above the ground must be placed at each corner, and on each end line center of the claim with all corners marked. Witness corner for inaccessible points.

(3) DISCOVERY WORK. A shaft, tunnel or open cut must be dug, the excavation to equal the work done to sink a shaft 4x6x8 feet deep, within 90 days in Arizona, Nevada, California, and New Mexico. Within 60 days in Wyoming and Colorado, one year in Utah.

(4) RECORD. File with the County Recorder a copy of the location notice within 90 days for a load claim and 60 days for a placer claim in Arizona, California, Colorado, Nevada, and within 60 days in New Mexico and Wyoming, and within 30 days in Utah.

PLACER CLAIM. Land containing mineral scattered or diffused and not in place may be located as a placer claim. Lodes discovered thereon should be filed on separately. 160 acres may be included in an association. (8 names at 20 acres each).

Building stone, sand, gravel, clays, pumice, limstone used for flux, gypsum, perlite, marble and salt beds may be acquired under the placer mining law.

Coal, oil, oil shale, sodium, phosphate, potash cannot be located. See the nearest United States Land Office for lease.

MILL SITE. A tract on non-mineral land of five acres may be located for mining or milling uses. It must be upon vacant land. No assessments.

TUNNEL CLAIM. Under Federal law a tunnel may be driven 3000 feet from the portal, along the tunnel site. The right to veins discovered relates back to the time of the location of the tunnel site.

PUBLIC LANDS—OIL & GAS. The law provides for prospecting permits and leasing. The acreage is limited to 2560 acres on any one geologic structure.

ASSESSMENT WORK. The Federal Law requires that to maintain title not less than one hundred dollars worth of labor shall be performed or improvements made during each year or until patent is acquired.

Example: A claim located September 1, 1954, and the discovery work completed within 90 days, annual work following should be completed between July 1, 1955 and July 1, 1956—Arizona, California, Colorado, Nevada, Wyoming.

Allowable expenditures include: Buildings, Machinery, Timber, Roadways, Trenching, Labor, Diamond drilling.

Federal law does not require recording of annual work done. Recording of such prevents attempted relocation.

APEX LAW. Extralateral rights allows the mining of ore from another's property. A vein may be followed downward on its dip beyond the sidelines of the claim. It cannot be followed beyond the end lines of the claim.

WHO MAY LOCATE A CLAIM. Any Citizen of the United States. A group of Citizens. A Corporation. No limit to the number of claims.

LANDS SUBJECT TO LOCATION. Vacant, unreserved, unappropriated. Surveyed or unsurveyed per land office. State land per State Land Commission. Railroad grant land per the railroads.

LAND WITHDRAWALS. All National Parks and Monuments unless there is an exception. Military Reservations. Power Sites per Federal Power Commission. Recreational Areas per Secretary of the Interior, Board of Supervisors or City authorities. Ranger Stations. Spring or water hole—miners should identify the 40 acre tract and apply to the U. S. Land Office—Public Water Reserve.

NATIONAL FORESTS. Mining is unrestriced, but consult Forest Rangers.

STATE GAME REFUGES. Mining is unrestricted.

GAZING DISTRICTS. Mining is unrestricted. District grazing offices should be consulted.

INDIAN RESERVATIONS. Unalloted land reserved for tribal Indians is subject to mineral lease by payment of royalties and annual lease rentals.

STATE LANDS. Arizona has sections 2, 16, 32 and 36. Other Western States have only two sections in each township. In some few instances the State may own the mineral rights on Federal land due to an exchange of land. Apply for a state lease.

RAILROAD LAND. Transcontinental railroads received land grants by acts of Congress. Mineral rights were usually excepted. The A. T. & S. F. Ry. is an exception. A lease must be obtained for mining on their land.

WATER RIGHTS. The Federal law recognizes the State laws as governing in the use of water. The owner of a mining claim is entitled to percolating waters and natural springs except certain mineral springs and a spring which does not constitute the source of a water course belongs to the owner of the land who map appropriate and use its entire flow. It is advisable to consult the State Land and Water Commissioners concerned with each source.

BEDS OF STREAMS. Minerals under navigable waters are the property of the State. The beds of unnavigable streams may be placer mined with temporary use of the water.

STOCK RAISING HOMESTEADS. The miner will be limited in his surface rights to the extent that he may be liable for damages. The Federal Government reserves the mineral rights.

BOULDER DAM RECREATIONAL AREA. Mining operations are allowed but with restrictions.

A.E.C. ALLOCATIONS OF MINING LEASES. Previous to leasing, potential uranium ground was withrawn for the purpose of exploration. After tests have been made by the Atomic Energy Commission, ore developed and costs of mining determined the ground is set up for leasing to competent, experienced mining men. These leases are subject to a directed ore clause. The Grand Junction Operation Office should be consulted.

TRANSFER OF URANIUM ORES. A license from the Atomic Energy Commission is needed to sell uranium and thorium ores.

DISCOVERY OF RADIOACTIVE ORE. A. E. C. engineers and geologists will assist prospectors in the field at the discovery and there is a free testing and assaying service.

PRIVATE LAND. Prospectors must come to an agreement with the owner of the property.

LANDS WITHDRAWN AND RESTORED. Effective on certain dates, bulletins and press releases will show by sections, townships, ranges and geographical areas the Tribal and Federal lands withdrawn for exploration or restored to the public domain. Contact United States Atomic Energy Commission—Grand Junction Operations Office, Colorado.

URANIUM OCCURRENCES
BY STATES AND COUNTIES

ARIZONA URANIUM . . . Occurrences by Counties

APACHE COUNTY: Lukachukai, Carrizo, Chilchinbito Area, Chinle Creek, Carrizo Mountains.

COCONINO COUNTY: Leupp, Young, Little Colorado, Grand Canyon, Marble Canyon, Vermillion Cliffs, Paria Plateau.

GILA COUNTY: Ellison Mining District.

MOHAVE COUNTY: Hack Canyon, Mt. Trumbull.

NAVAJO COUNTY: Cameron—Holbrook Areas, Kayenta, Moonlight Creek.

PIMA COUNTY: Ruby—Oro Blanco Mining district.

PINAL COUNTY: Ray.

CALIFORNIA URANIUM . . . Occurrences by Counties

KERN COUNTY: Bodfish, Glenville, Upper Jawbone Canyon, Summit, Rosamond, Searles Station.

SAN BERNARDINO COUNTY: Clark Mountain, Twenty-nine Palms.

VENTURA COUNTY: Piru Creek.

COLORADO URANIUM . . . Occurrences by Counties

BOULDER COUNTY: Eldora, Nederland, Caribou mine area.

CUSTER COUNTY: Silver Cliff.

DELTA COUNTY: Gunnison area near Delta.

DOLORES COUNTY: Slick Rock, Dove Creek, Rico.

FELLER COUNTY: Cripple Creek.

FREMONT COUNTY: Canon City, Radiant, Penrose area.

GARFIELD COUNTY: Rifle, Newcastle.

GILPIN COUNTY: Central City.

GRAND COUNTY: Sulphur Springs, Troublesome.

GUNNISON COUNTY: Gunnison area.

HINSDALE COUNTY: Eureka District.

JEFFERSON COUNTY: Golden area.

LAKE COUNTY: Leadville District.

LA PLATA COUNTY: Durango area.

LARIMER COUNTY: Manhattan Mining District, and area in Township 9 North, Range 70 West.

MESA COUNTY: Gateway, Salt Wash.

MOFFAT COUNTY: Red Wash, Massadona area.

MONTROSE COUNTY: Uravan, Naturita, Paradox, Bull Canyon.

OURAY COUNTY: Vicinity of Ouray.

PITKIN COUNTY: Aspen District.

RIO BLANCO: Meeker area, Marvin.

ROUTT COUNTY: Fish Creek area.

SAN JUAN COUNTY: Rico District.

SAN MIGUEL COUNTY: Gypsum Valley, Bull Canyon, North of Egnar, Vanadium, Telluride.

dAtio UureNewfiYJAN

NEVADA URANIUM . . . Occurrences by Counties

CLARK COUNTY: Overton—Logan mining district.

ESMERALDA COUNTY: Blair, Coaldale.

HUMBOLDT COUNTY: Oreana — near Star mining district, Dacia Creek, Virgin Valley—Township 45 North, Range 26 East.

LANDER COUNTY: Austin mining district to the south.

MINERAL COUNTY: Rhodes.

NYE COUNTY: Belmont.

WHITE PINE COUNTY: Hunter mining district.

NEW MEXICO URANIUM . . . Occurrences by Counties

BERNALILLO COUNTY: Albuquerque area.

GRANT COUNTY: White Signal District.

LEA COUNTY: Hobbs.

LINCOLN COUNTY: White Oaks.

McKINLEY COUNTY: Haystack Butte, Laguna, Grants.

SAN JUAN COUNTY: Shiprock area.

SANTA FE COUNTY: West of Golden, Cerrilos mining district.

SIERRA COUNTY: Elgle, Hillsboro, Pittsburg mining district.

COCORRO COUNTY: Magdalena.

UTAH URANIUM . . . Occurrences by Counties

BEAVER COUNTY: Star mining district.

EMERY COUNTY: Temple Mountain, San Rafael Swell, Greenriver district, Muddy River.

GARFIELD COUNTY: Circle Cliffs, Water Pocket Fold, Escalante district, Henry Mountains, Henrieville, Hite, Trachyte Creek.

GRAND COUNTY: Thompsons, Basin, Moab, Dolores River at Beaver Creek, Onion Creek, Richardson.

JUAB COUNTY: Fish Springs, Tintic.

KANE COUNTY: Pahrea, Orderville Gulch at Township 40 South, Range 9 West, Kaiparowits Plateau.

PIUTE COUNTY: Marysvale.

SAN LUAN COUNTY: La Sal District, Steen Mine, Abajo, Cane Canyon, Upheaval Dome, Big Indian, Monticello, Blanding, Bluff, Navajo Mountain, Natural Bridges Monument, White Canyon.

TOOELE COUNTY: Ironton.

UINTAH COUNTY: Vernal, Myton, Ouray, Cliff Creek, Bad Land Creek.

WASHINGTON COUNTY: Leeds, Harrison.

WAYNE COUNTY: Torrey, Cainville, Hankville, Browns Ranch, Dirty Devil River.

WYOMING URANIUM . . . Occurrences by Counties

CAMPBELL COUNTY: Gillette, Savageton, Pine Tree, Pine Tree Junction, Pumpkin Buttes, Township 42 North, Range 76 West.

CARBON COUNTY: Baggs, Encampment, Saratoga, Seminoe Mountains, Wamsutter.

COWLEY COUNTY: Two miles southwest of Cowley near Lovell, Shell Creek about 15 miles northwest of Graybull, Section 3, Township 53 north, Range 94 West. Section 26, Township 55 north, Range 95 west, Section 10, Township 50 north, Range 90 west, Section 34, Township 51 north, Range 90 west.

CROOK COUNTY: Belle Fourche River area—Beulah, Hulett, Aladdin.

FREEMONT COUNTY: Crooks Gap, Green Mountains, East of Lenore, Moneta.

HOT SPRINGS COUNTY: Grass Creek, Section 16, Township 46 north, Range 98 west.

JOHNSON COUNTY: Township 42 north, Range 77 west, East of Kaycee, Northeast of Midwest.

NATRONA COUNTY: Hiland.

SWEETWATER COUNTY: Township 26 north, Range 94 west.

VARIOUS GEOLOGICAL AREAS CORRELATED

Economic ores range between 0.1% and 0.4%. Uranium occurs in most granites where it is concentrated in the minerals such as monazite, xenotime and zircon in amounts from .0003% to .0007%. Pegmatite dikes furnish more than the granites but the grade is low. Granitic Magmas seem to be the primary source of uranium.

In the Belgian Congo, Shinkolobwe Mine pitchblende bearing veins are associated with molybenite, tourmaline and monazite in Pre-Cambrian quartz and shales.

In Czechoslovakia and in Portugal, pitchblende veins of the Paleozoic Era are in granitic batholiths.

At Cornwall, England, pitchblende is associated with copper ores.

At Great Bear Lake, Canada, the Eldorado Mine and the Gunnar Mine produce pitchblende from veins in sedimentary and volcanic rocks.

At Montreal River on the east shore of Lake Superior, pitchblende occurs along diabase dikes which cut granite-gneiss of Pre Cambrian age.

In the Colorado Front Range, pitchblende deposits are in veins and fissures of Pre-Cambrian schists and granite-gneisses.

At Marysvale, Utah, pitchblende occurs below autunite and meta-torbernite in veins. Some of the deposits are in Tertiary quartz monzonite and rhyolite.

In Western Montana the uraninite minerals pitchblende, gummite, autumite, torbernite, zeunerite and uranophane occupy reefs and are associated with pyrite, galena, molybdenite, ruby silver and chalcedonic quartz. The Coeur d' Alene Idaho pitchblende is associated with silver, hematite and quartzite of the Pre-Cambrian Era. Some shales of Sweden contain bitumen bearing uranium up to 0.5%. The richest known shale beds are of the Mesozoic Era. It has been suggested that shales such as these are in due part to uranium-bearing bacteria. Marine phosphorites of Idaho, Wyoming, Utah and Montana of the palezoic - Permian period show light concentrations of uranium. In Florida the phosphorites with uranium are of the Cenozoic - Pliocene period. The uranium limestone beds of New Mexico are in the Mesozoic - Jurassic period — Todilto formation. The ore minerals are pitchblende, carnotite, tyuyamunite, uranophane, pyrite and fluorite. The Witwatersrand reefs of South Africa contain pitchblende associated with gold in quartz conglomerate. In the Rhodesian area pitchblende has been found with copper. South Australia - Radium Hill area produces uranium associated with davidite and iron - titanium mineral in granite and amphibolites. The Rum Jungle area near Darwin is of the pre-Cambrian Era. The copper-uranium uraninite is associated with pyrite and chalcopyrite with slate. The Black Hills area of South Dakota and Wyoming produce ores of the Mesozoic - Cretaceous - Dakota and Lakota formations. Utah produces uraninite in the Cenozoic Era - Tertiary period rocks of igneous origin at Marysvale. Colorado and Utah produce up to 20% uraninite from the Mesozoic Era - Jurassic period — Shinarump Conglomerate.

PROSPECTING FOR URANIUM

When prospecting for uranium in the Colorado River Pateau Areas —"The Four Corners"—formations will be dealt with which are principally sandstone.

If the search is carried on along the Continental Divide the same formations that contain silver, copper, vanadium and other minerals should be looked over and vein systems will be studied.

Those areas many miles from either the Colorado River Plateaus or the Continental Divide should be checked around mines that have produced silver, copper, cobalt, iron, lead and vanadium. Check areas showing evidence of hot springs, fluorspar, phosphate, alunite and some of the coal and asphaltic provinces.

Study the mineral areas shown on the map then by the use of geologic information correlate given areas. That is, a material of a certain geologic age for instance palezoic, may be found in New Mexico and also in California thus affording a chance of finding similar kinds of minerals in the two places.

The geologic study will familiarize the prospector with the terms used for a differentiation of the several layers of sedimentary beds of the Colorado Pateau.

These beds are horizontal or tilted except where disturbed by anticlines, folds, faults or by igneous intrusions. Prospect around these disturbed areas. Of the many sedimentary beds that range in age from Palezoic to Tertiary there are fourteen known formations which contain uranium.

The uranium deposits in the colorado Plateau are bedded deposits in sandstone, limestone and shales and in general parallel to the bedding. Geographically these deposits cover a wide area. From Greenriver, Utah they extend to Holbrook in Arizona a distance roughly of three hundred miles; from Skull Creek Colorado to Grants, New Mexico; and from Rifle Colorado to Leeds in southwestern Utah.

It is possible that a broad uranium province will be found somewhere in the western states. The recent evidence supporting the hydrothermal theory, and the discovery of bacteria producing radio activity which calls for the study and search for the unknown elements responsible for uranium, adds weight to the supposition that uranium will be found in larger bodies and more types of material and the possibility that all of the known uranium ores may have come from one source.

Some of the principle formation types with the kind of ore follows:

(1) Entrata—Roscoelite. High Vanadium—low uranium.
(2) Morrison—Yellow Carnotite. Copper, Selenium, silver.
(3) Shinarump Temple Mountain—High uranium—low vanadium with selenium, copper, arsenic, cobalt and iron.
(4) Shinarump Conglomerate—High Carnotite.
(5) Chinle—Yellow Carnotite with blue-green copper and silver stains. Leeds, Utah.
(6) Jurassic—Todilto Limestone—Carnotite, uranophane and tuyuyamunite. Grants, New Mexico.
(7) Jurassic Asphaltite deposits—Carnotite associated with vanadium. Found in Utah, California and Arizona.
(8) Miocene—Rosamond—Autunite. California. Generally whether in beds or veins the uranium ores are irregularly distributed and in relatively small deposits.

IDENTIFICATION TESTS

Most of the uranium minerals are brightly colored, yellow or green and the rich pitchblend and uraninite ores are of a distinctive black, grayish color, perhaps with copper and iron stains. These ore bodies are recognized in outcrops but this seldom occurs. Careful study of geological and mineralogical conditions together with the use of portable "counters" to test the radioactive ores is in order.

There are many good portable counters on the market which range in price from $25. to $2,000. The lower priced Geiger Counter is very popular as is the also higher priced Scintillation Counters.

Carnotite may be identified when the ore fuses easily to a black mass containing small pellets, when it yields water if heated in a tube closed at one end, when it dissolves in hot dilute sulfuric acid and yields a yellow or orange solution that turns brown when hydrogen peroxide is added. It yields a yellow solution in hydrocloric acid and a green solution in nitric acid. Being radio active it can be tested with an electroscope, radioscope, Geiger counter or scintillation instruments.

(97)

MAP

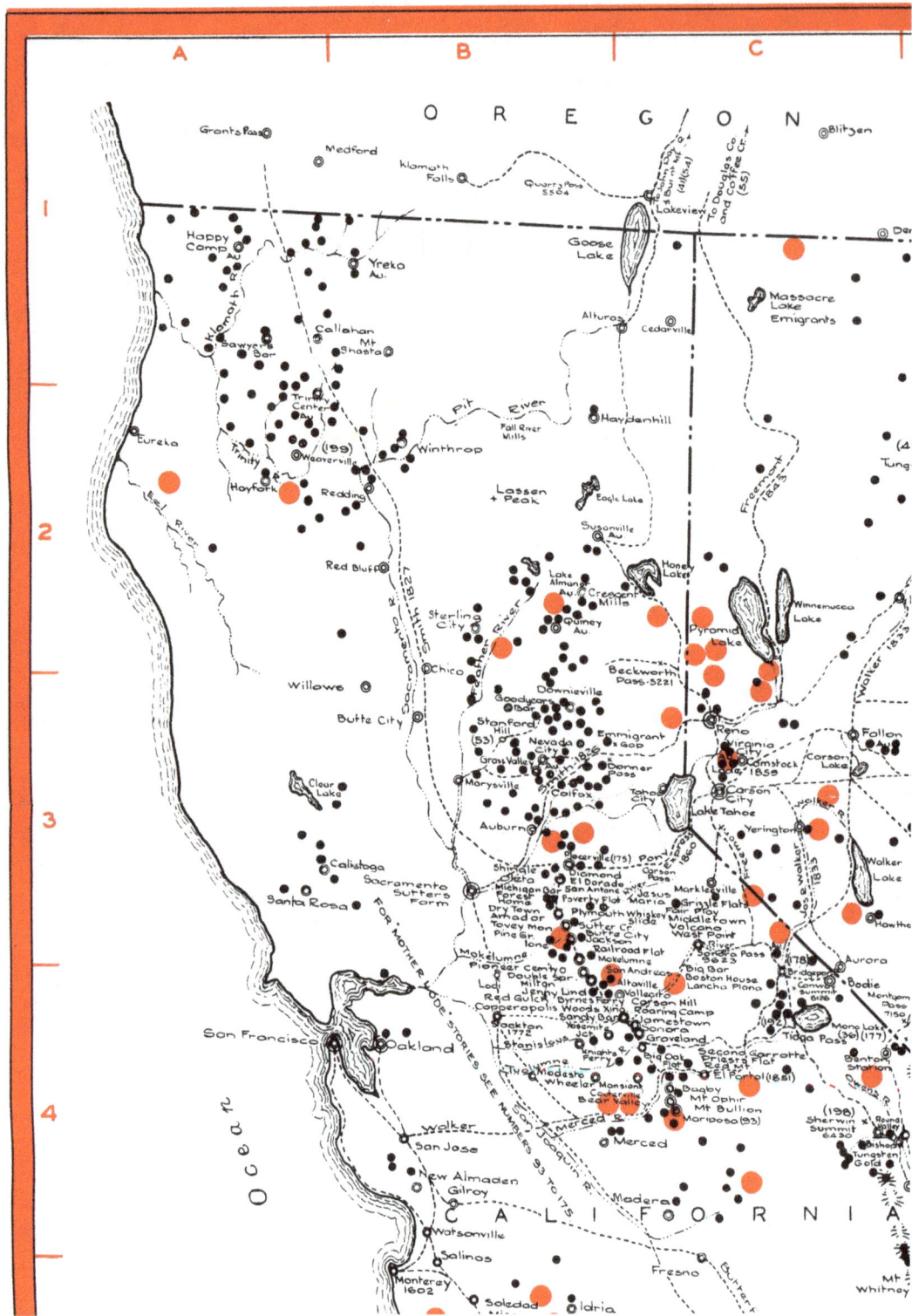

A   B   C

O R E G O N

Grants Pass
Medford
Klamath Falls
Quartz Pass 5504
To John Day
& Burnt Mt (all/s.a.)
To Douglas Co. and Coffee Cr. (55)
@Blitzen
@Der

Happy Camp
Yreka Au.
Goose Lake
Lakeview
Massacre Lake Emigrants

Callahan Mt Shasta
Alturas
Cedarville

Sawyers Bar
Trinity Center Au.
Pit River
Fall River Mills
Haydenhill
Fremont 1843
(4
Tung

Eureka
Weaverville
(99)
Winthrop
Lassen + Peak
Eagle Lake

Hayfork
Redding
Trinity

Red Bluff
Susanville Au.
Honey Lake

Lake Almanor Au.
Crescent Mills
Sterling City
Quincy Au.
Winnemucca Lake

Willows
Chico
Downieville
Beckworth Pass 5221
Pyramid Lake
Reno
Fallon

Butte City
Goodyears Bar
Stanford Hill
(53)
Nevada City
Grass Valley
Emigrant Gap
Virginia City
Comstock 1859
Carson Lake

Clear Lake
Marysville
Donner Pass
Tahoe
Carson City
Lake Tahoe

Calistoga
Colfax
Auburn
Yerington
Walker Lake

Sacramento Sutters Farm
Placerville (175)
Port
Carson Pass
Hawtho

Santa Rosa
Shingle Diamond
Michigan Bar
El Dorado
San Antone
Jesus Maria
Markleville
Grizzle Flats

Forest
Poverty Flat
Hair Dry
Dry Town
Plymouth Whiskey
Middletown
Volcano
West Point

Armador
Sutter Cr.
Pine Gr.
Tovey Mon.
Ione
Jackson City
Railroad Flat
Mokelumne
Sonora Pass
(178)
Aurora
Bridgeport
Bodie

San Francisco
Oakland
Mokelume
Pioneer Cemtry
Double Spr.
San Andreas
Altaville
Big Bar
Boston House
Lancha Plana
Summit 8488
Montym Pass 7150

Lodi
Emry Lind
Byrnes Ferry
Copperapolis 1772
Red Gulch
Woods Xing
Vallecito
Carson Hill
Jamestown
Sonora
Tioga Pass
Mono Lake

Stockton
Roaring Camp
Sandy Bar
Yosemit
Groveland
Second Garrotte
Priests Flat
(178)

Stanislaus
Knights
Big Oak Flat
El Portal (1851)
Santa Station

Two Modesto
Wheeler Mansions
Coultervile
El Portal
Bugby
Mt Ophir
(198)
Sherwin Summit 8430
Round
Tungsten Gold

Walker
Merced S
Bear Valle
Mt Bullion
Mariposo (53)

San Jose
Madera
Merced

New Almaden
Gilroy
C A L I F O R N I A

Watsonville
Salinas
Fresno
Buttort
Mt Whitney

Monterey 1602
Soledad
Idria

Ocean
FOR MOTHER LODE STORIES SEE NOS.
NUMBER 93 TO 175
San Joaquin R.
1

1
2
3
4

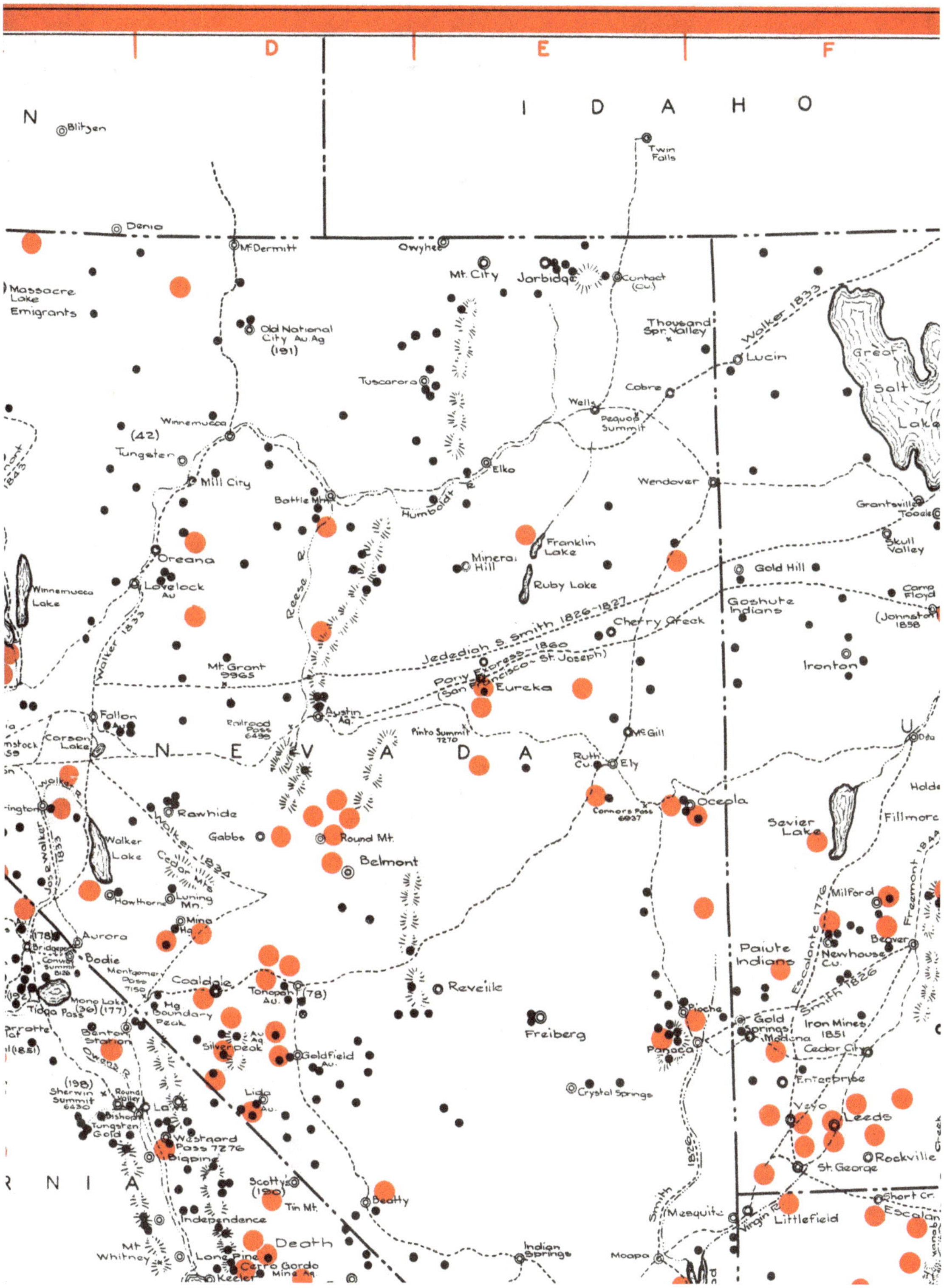

N

I D A H O

Blitzen

Twin Falls

Denio

McDermitt
Owyhee
Mt. City
Jarbidge
Contact (Cu)

Massacre Lake
Emigrants

Old National City Au Ag (191)

Thousand Spr. Valley

Walker 1833

Lucin

Great Salt Lake

Tuscarora

Wells
Cobre
Pequop Summit

Winnemucca
(42)
Tungsten

Elko

Wendover

Grantsville
Tooele

Mill City

Battle Mtn.

Humboldt

Skull Valley

Oreana

Mineral Hill

Franklin Lake

Gold Hill

Camp Floyd

Lovelock Au

Ruby Lake

Goshute Indians

(Johnston 1858)

Winnemucca Lake

Reese R.

Cherry Creek

Jedediah S. Smith 1826-1827

Ironton

Mt. Grant 5965

Pony Express 1860
(San Francisco - St. Joseph)

Eureka

Fallon

Railroad Pass 6490

Austin Au

McGill

U
Deu

Carson Lake
Comstock

N E V A D A

Pinto Summit 7270

Ruth Cu
Ely

Osceola

Holde

Walker

Walker 1834

Rawhide
Gabbs
Round Mt.

Connors Pass 6937

Sevier Lake

Fillmore

Walker Lake
Cedar Mtn.

Belmont

Milford

Luning
Mina
Mt.

Hawthorne

Freemont 1844

Aurora
Bridgeport
Conway Summit 8138

Mine
Paiute Indians

Newhouse Cu

Beaver

Bodie
Montgomery Pass 7150

Coaldale

Reveille

Escalante 1776

Mono Lake (177)
Tioga Pass (20)

Mg
Boundary Peak

Tonopah (78)

Freiberg

Gold Springs
Modena 1851

Iron Mines

Smith 1826

Benton Station
Owens

Silver Peak

Goldfield Au

Panaca
Pioche

Cedar City

Parrotte Gr (1881)

(198)
Sherwin Summit 6820
Round Valley

Lida Au

Crystal Springs

Enterprise

Virgin R.

Bishop
Tungsten Gold

La

Vaya
Leeds

Westgard Pass 7276
Bigpine

Scotty's (190)

Rockville

St. George

Death

N I A

Mt. Whitney

Deatty
Tin Mt.

Mesquite
Littlefield

Escala

Independence

Indian Springs

Short Cr.

Long Pine
Cerro Gordo Mine Ag

Keeler

Moapa

Md

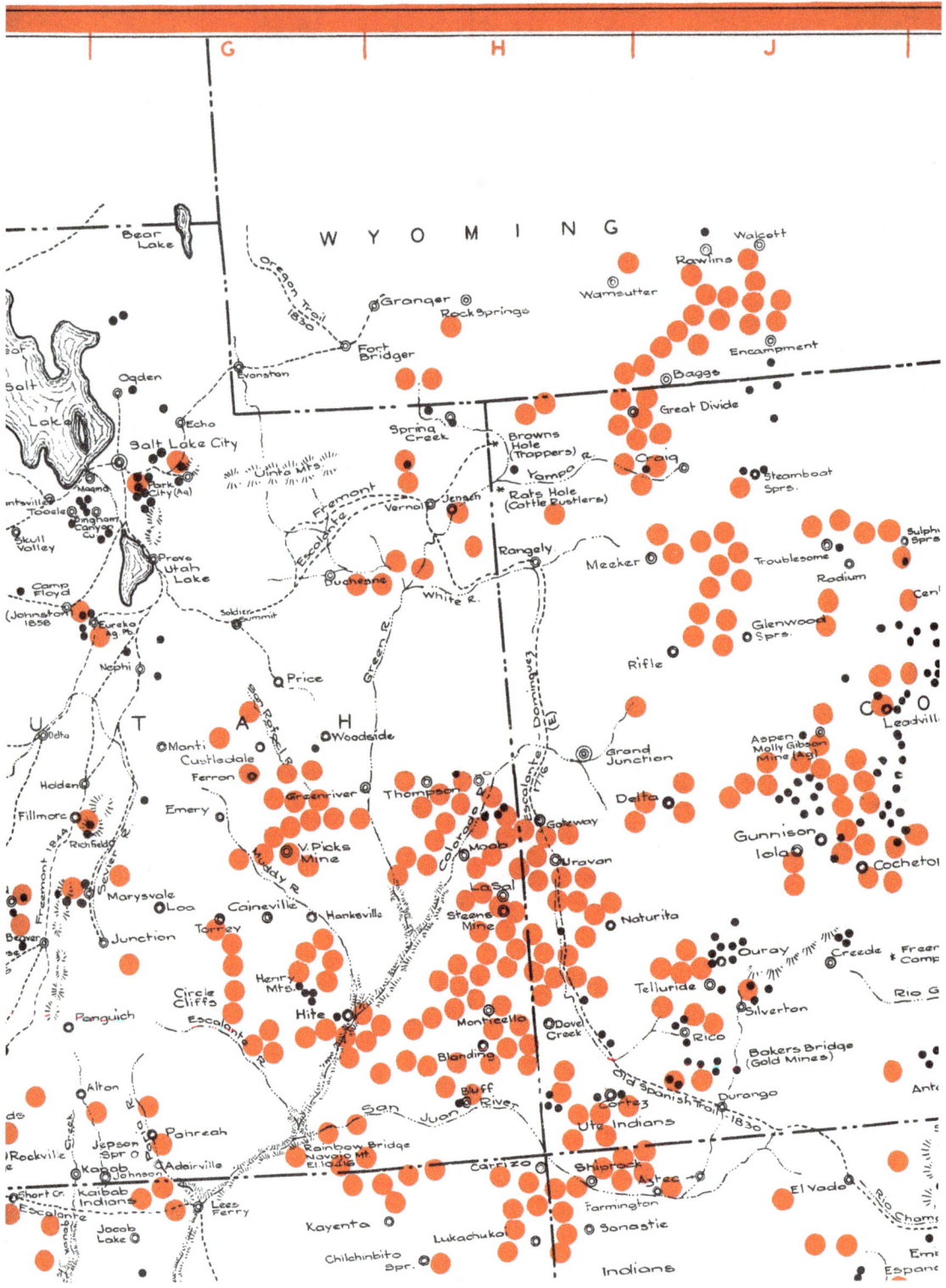

WYOMING

Bear
Lake

Oregon Trail 1830

Granger
Rock Springs

Warnsutter
Rawlins
Walcott

Baggs

Encampment

Fort
Bridger

Evanston

Ogden

Echo

Salt
Lake

Salt Lake City

Park City (Ag)

Magna

Bingham Canyon Cu

Tooele

Skull
Valley

Camp
Floyd
(Johnston
1858)

Eureka Ag Pb

Nephi

Spring
Creek

Uinta Mts.

Fremont Escalante

Vernal

Jensen

Browns
Hole
(Trappers)

Yampa

Rats Hole
(Cattle Rustlers)

Great Divide

Craig

Steamboat
Sprs.

Rangely

Meeker

Troublesome

Radium

Cent

Sulph
Spra

Duchesne

White R.

Soldier Summit

Green R.

Price

U T A H

Manti

Castledale

Ferron

Emery

Woodside

Greenriver

Thompson

Escalante-Dominguez

Grand
Junction

Delta

Glenwood
Sprs.

Rifle

C O L O

Leadvill

Aspen
Molly Gibson
Mine (Ag)

Gunnison
Iola

Cochetoi

Holden

Fillmore

Richfield

Sevier

Fremont 1844

Marysvale

Loa

Caineville

Hanksville

Muddy R.

V. Picks
Mine

Colorado R.

Moab

Gateway

Uravan

Naturita

La Sal

Steens
Mine

Ouray

Creede

Freer
Comp

Rio G

Beaver

Junction

Torrey

Circle
Cliffs

Henry
Mts.

Hite

Escalante

Panguich

Monticello

Blanding

Dove
Creek

Rico

Telluride

Silverton

Bakers Bridge
(Gold Mines)

Durango

Ante

Alton

Jepson
Spr

Pohreah

Rockville

Kanab

Johnson

Adairville

Rainbow Bridge
Navajo Mt
El 10.416

San Juan River

Bluff

Carrizo

Ute Indians

Shiprock

Aztec

El Vado

Rio Cham

Short Cr

Escalante

Kaibab
Indians

wanabb

Lees
Ferry

Jacob
Lake

Kayenta

Chilchinbito
Spr.

Lukachukai

Sonastie

Farmington

Indians

Emn

Espano

K   L   M

NOTE: This map has been drawn for use with McAllister's Lost Mines, Fabulous Uranium and Buried Treasures of the Southwest

INDEX OF PLACES MENTIONED IN McALLISTER'S "LOST MINES"

NUMBERS AND LETTERS IN PARENTHESES REFER TO THE PARAGRAPHS IN THE BOOK

(A)—7K Cabeza De Vaca
(B)—6J "Francisco Vasques De Coronado"
(C)—8L "Don Antonio De Espejo"
(D)—8K "Juan De Onate"
(E)—3H "Escalante and Dominquez"
(F)—8O "Pedro Vial"
(G)—4L "Zebulon Pike"
(H)— "Josiah Gregg"
(J)—7F "Butterfield-Overland Mail"
(K)—7E "Sonora Trail"
(1)—10J "Aztec Treasure"
(2)—10J "Mesa of the Bulls"
(3)—10H "Treasure of Bachaca"
(4)—7F "Two Suns East"
(5)—6F "Lost Nugget Mine"
(6)—6F "Lost Nugget Mine"
(7)—6F "Caugh Oir Golden Cup"
(8)—10J "Treasure of Don Felipe"
(9)—5E "Lost Breyfogle Mine"
(10)—7F "Lost Sopon Mine"
(11)—10H "Lost Tayope Mine"
(12)—10J "Lluvia De Oro"
(13)—7G "Lost Escalante Mine"
(14)—7F "Lost Soapmaker Mine"
(15)—6F "Lost Six-Shooter Mine"
(16)—6F "The Glory Hole"
(17)—10J "The Golden Beans"
(18)—6E "Black Gold"
(19)—7G "Good Medicine and Burried Gold"
(20)—8H "The Copper Box"
(21)—8H "Lost Mine and Burried Treasure of the Tumacacori Mission"
(22)—9H "Buried Gold of Casa Grande
(23)—7G "Red Rock Treasure"
(24)—10K "Maxmilian's Gold"
(25)—7D "Lost Peg Leg Mine"
(26)—7F "Brady-Mines"
(27)—6F "Lost Squaw Mine"
(28)—6G "Legend of Dr. Thorne"
(29)—6H "Mines of New Mexico"
(30)—7G "Planchas de Plata"
(31)—7H "Lost Yuma Mine"
(32)—6G "Lost Squaw Hollow Mine"
(33)—6H "Lost Adams Diggings"
(34)—6F "Lost Mine of Don Miguel Peralta"
(35)—6G "Lost Dutchman Mine"
(36)—6G "Lost Cement Mine"
(37)—6G "Lost Shoemaker Placer"
(38)—7G "Montezuma's Treasure"
(39)—8G "Shepherd's Lost Mine"
(40)—1C "Penhachape Mine"
(41)—1C "Lost Blue Bucket Placer"
(42)—2D "Lost Mine of the Little Brown Men"
(43)—7M "Geronimos Mine"
(44)—7E "Lost Cowboy Mine"
(45)—6M "Silver Moun lain"
(46)—5H "The Lost Door Mine"
(47)—7E "Lost Bandit Mine"
(48)—6G "Waggoner's Lost Ledge"
(49)—6E "Lost Arch Mine"
(50)—6G "Lost Mines of the Peraltas"
(51) 6G "Pedro Peralta's Mine"
(52)—6E "Lost Dutch Oven"
(53)—2D "Lost Indian Mine of Butte County"
(54)—1C "Blue Bucket Placer"
(55)—1C "Ed Schieffelin's Tombstone Ledge"
(56)—7H "Silver King Mine"
(57)—6G "The Lost Soldier's Mine"
(58)—7N "The San Saba"
(59)—8N "The Gold Mines of the Nueces"
(60)—10N "The Rock Pens"

(61)
(62)
(63)
(64)
(65)
(66)
(67)
(68)
(69)
(70)
(71)
(72)
(73)
(74)
(75)
(76)
(77)
(78)
(79)
(80)
(81)
(82)—
(83)—
(84)—
(85)—
(86)—
(87)—
(88)—
(89)—
(90)—
(91)—
(92)—
(93)—
(94)—4
(95)—4
(96)—4
(97)—4
(98)—4
(99)—4
(100)—4
(101)—4
(102)—4
(103)—4
(104)—4
(105)—4
(106)—4
(107)—4
(108)—4
(109)—4
(110)—4
(111)—4
(112)—4
(113)—4
(114)—4

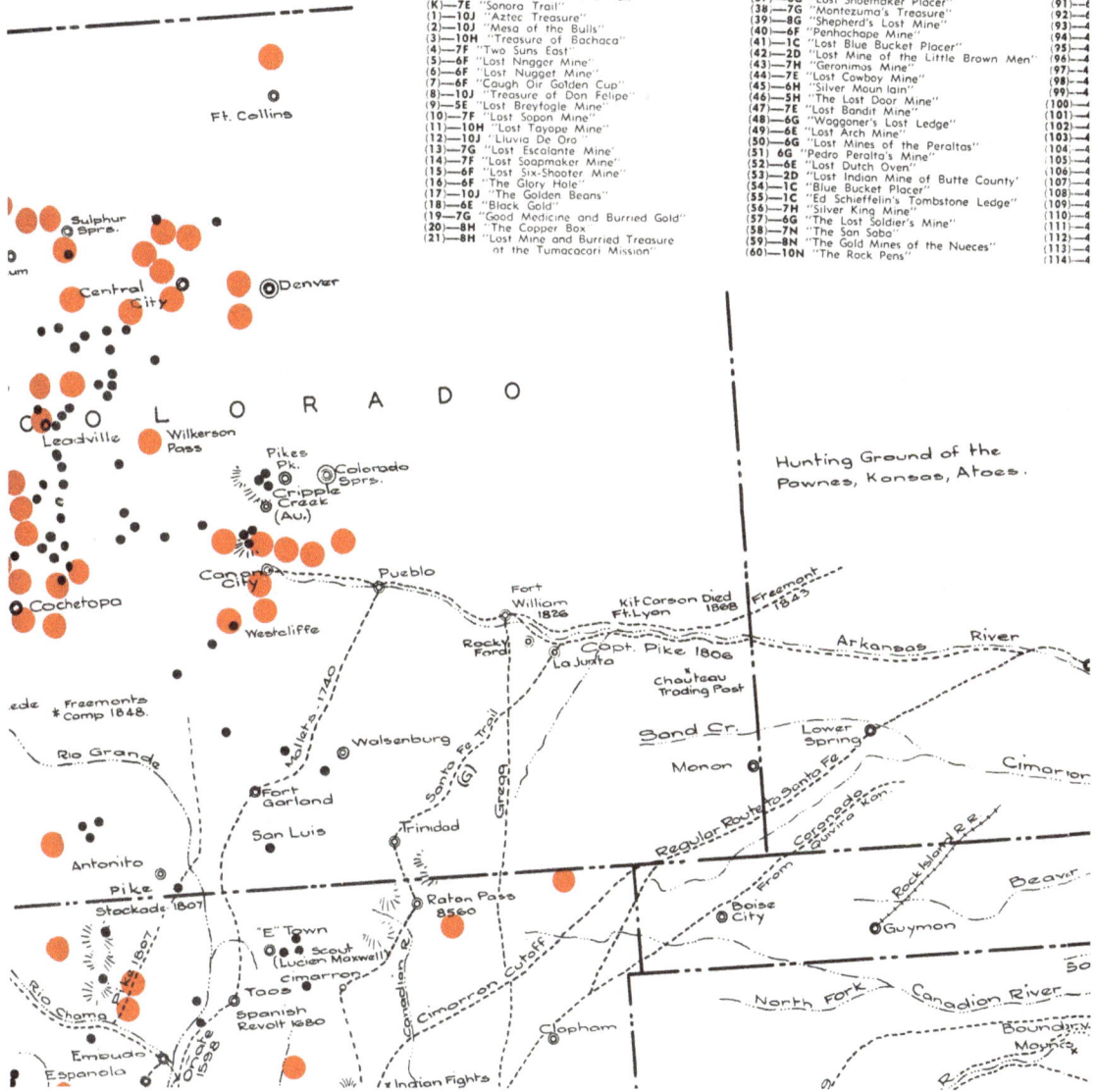

Ft. Collins

Sulphur Sprs.

Central City

Denver

C O L O R A D O

Leadville

Wilkerson Pass

Pikes Pk.

Colorado Sprs.

Cripple Creek (Au.)

Canon City

Cochetopa

Westcliffe

Pueblo

Hunting Ground of the Pawnes, Kansas, Atoes.

Fort William 1826

Kit Carson Died Ft.Lyon 1868

Freemont 1843

Capt. Pike 1806

Arkansas River

Rocky Ford

La Junta

Chauteau Trading Post

ede Freemonts Comp 1848.

Rio Grande

Walsenburg

Sand Cr.

Lower Spring

Cimarron

Monon

Santa Fe Trail

Regular Route to Santa Fe

Coronade

From

Fort Garland

San Luis

Trinidad

Beaver

Antonito

Pike Stockade 1807

Raton Pass 8560

Boise City

Guymon

"E" Town

Scout (Lucien Maxwell)

Cimarron

Toos

Spanish Revolt 1680

North Fork

Canadian River

50

Rio Chama

Canadian R.

Cimarron Cutoff

Clapham

Boundary

Moynes

Embuds

Espanola

Onate 1598

Indian Fights

NOTE: This map has been drawn for use with McAllister's Lost Mines, Fabulous Uranium and Buried Treasures of the Southwest

INDEX OF PLACES MENTIONED IN
McALLISTER'S "LOST MINES"

NUMBERS AND LETTERS IN PARENTHESES
REFER TO THE PARAGRAPHS IN THE BOOK

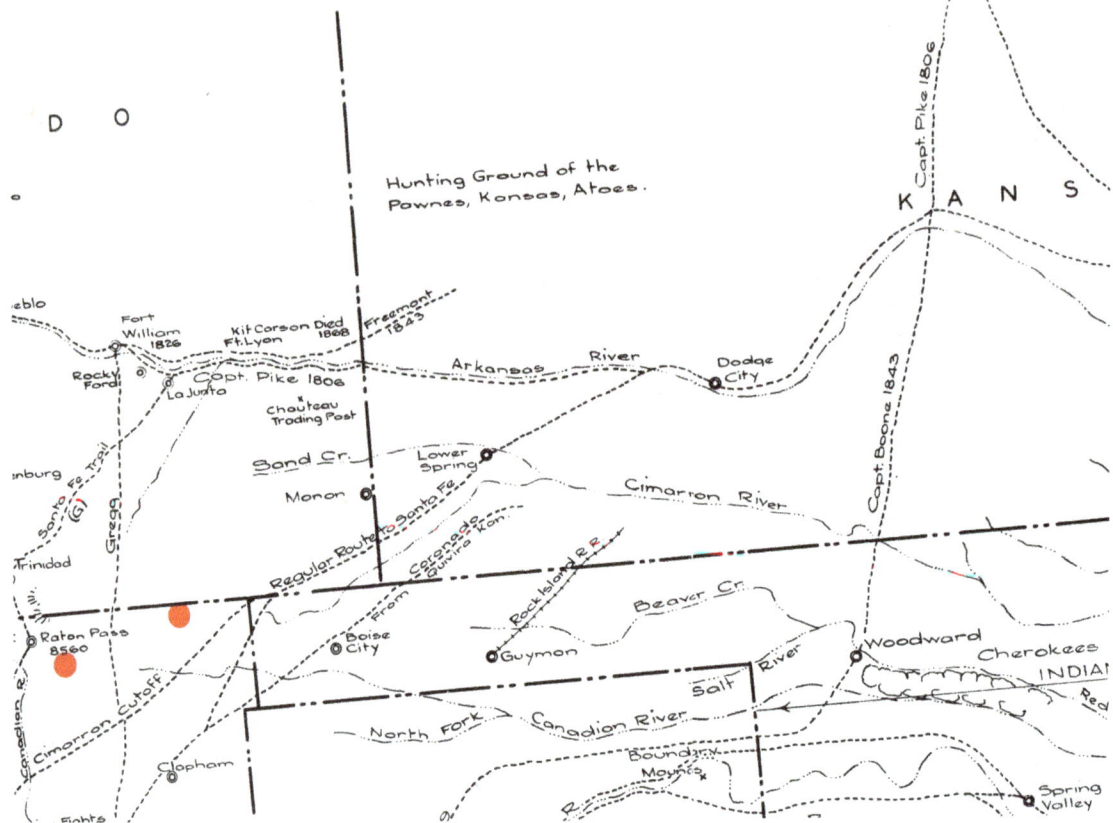

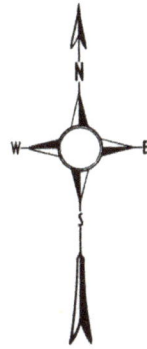

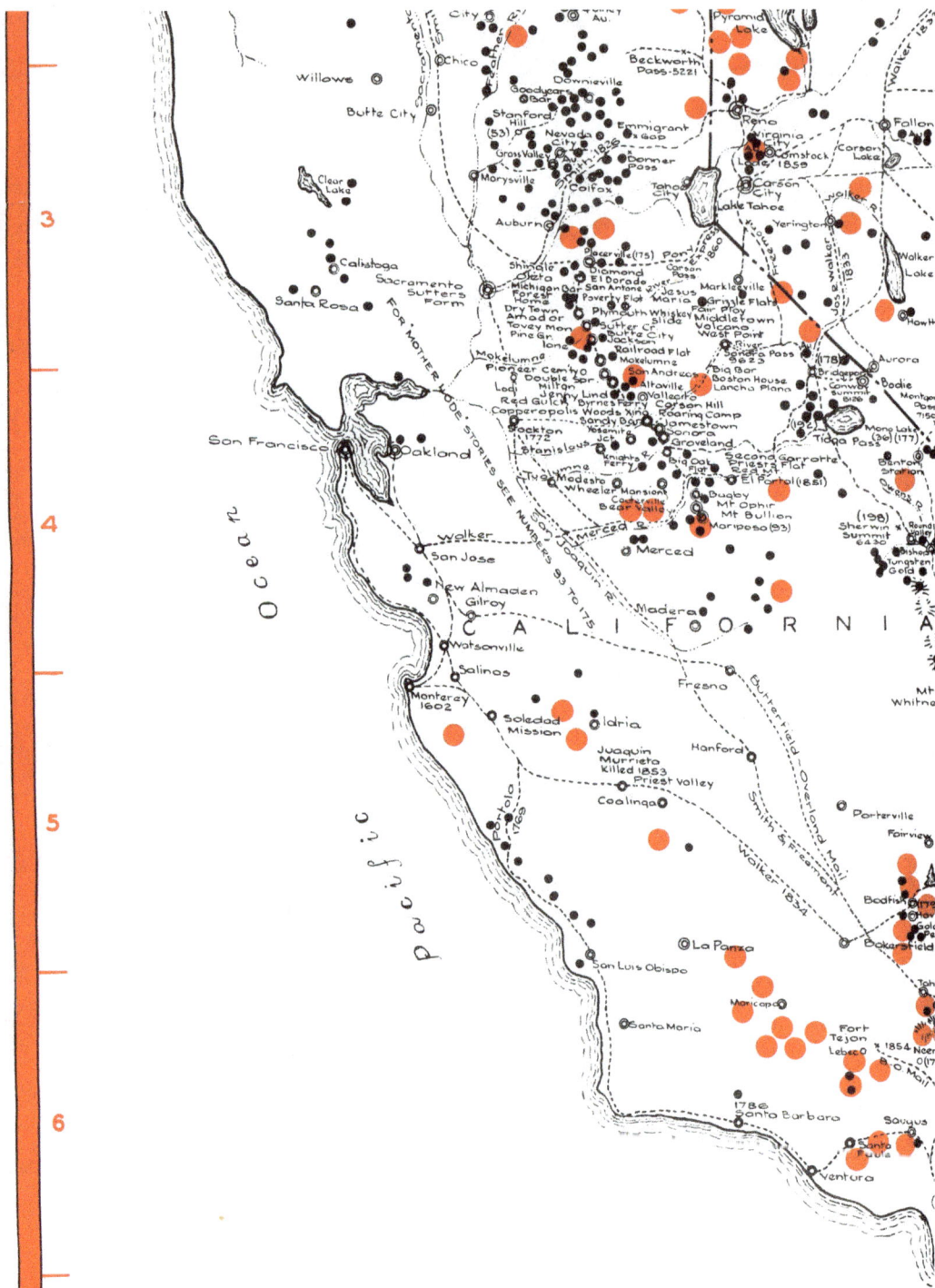

URANIUM PROSPECTING

When prospecting for uranium in the Colorado River Plateau Areas —"The Four Corners"— formations will be dealt with which are principally

Some of the principal formation types with the kind of ore follows:

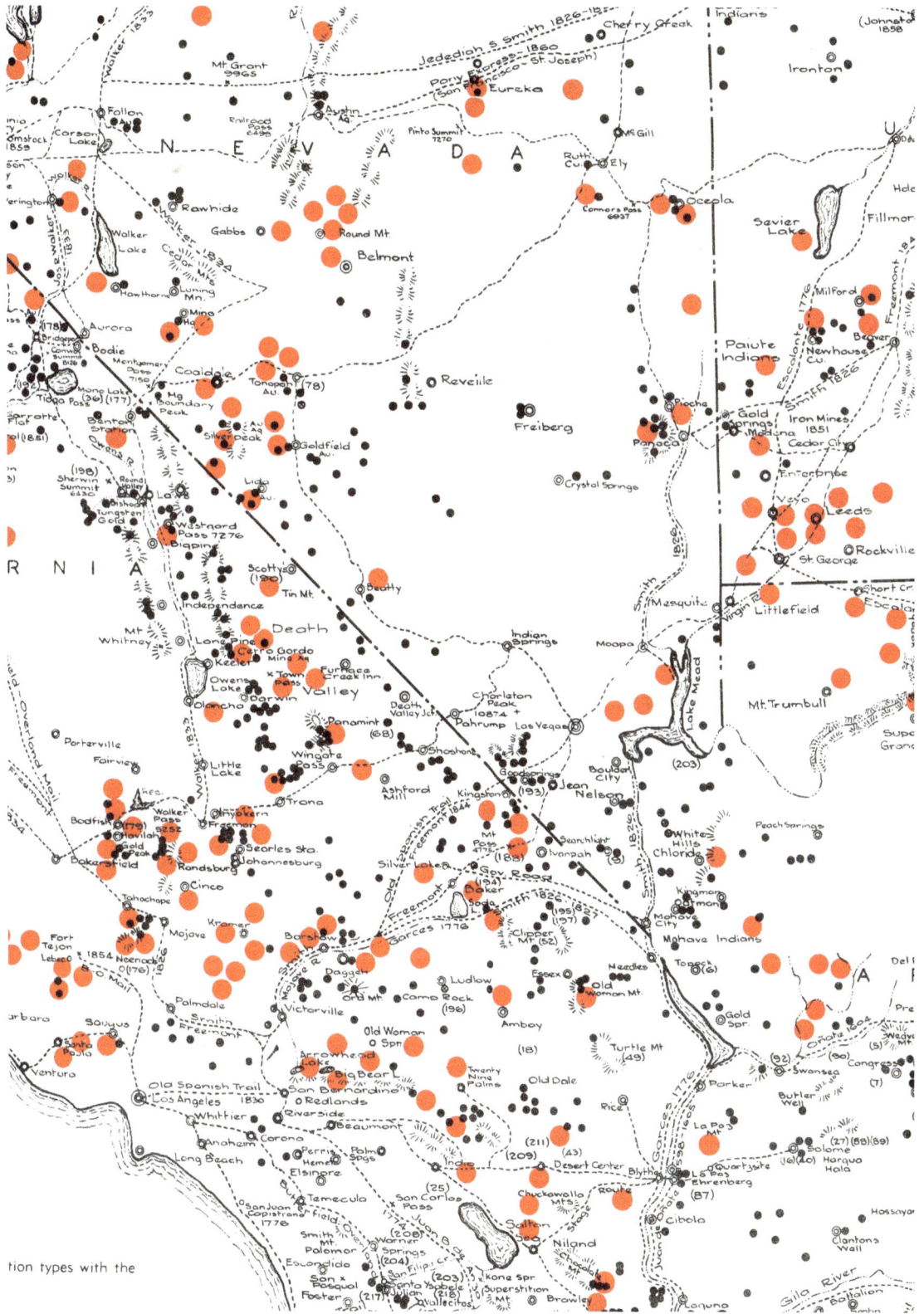

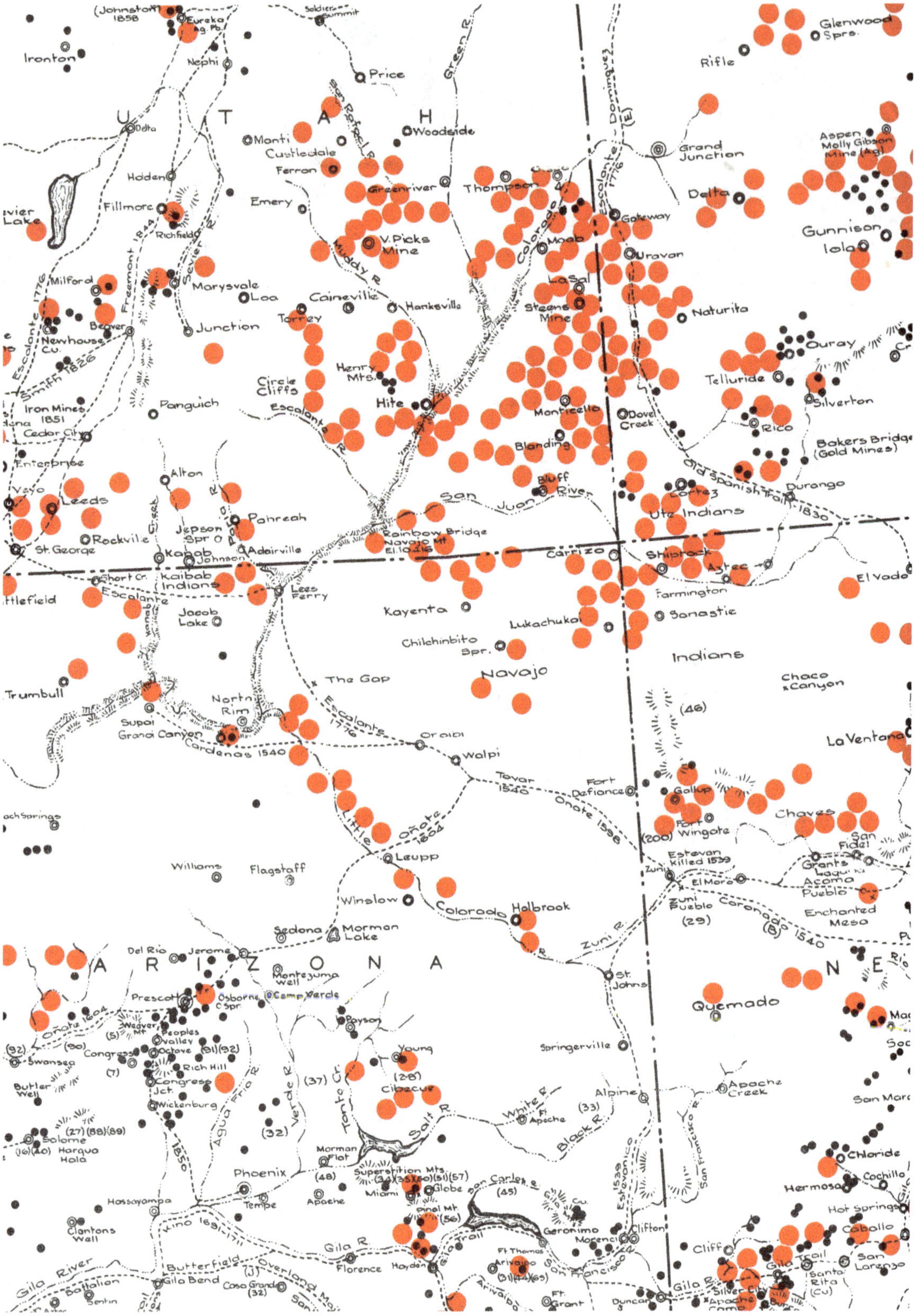
U T A H
ARIZONA
N E
(Johnston) 1858
Eureka Ag. Pb.
Soldier Summit
Ironton
Nephi
Price
Green R.
Rifle
Glenwood Sprs.
Delta
San Rafael R.
Woodside
Monti
Castledale
Ferron
Grand Junction
Aspen
Molly Gibson Mine (Ag)
Holden
Emery
Greenriver
Thompson
Escalante 1776
Dominguez
Delta
Sevier Lake
Fillmore
Richfield
Fremont 1842
Sevier R.
V. Picks Mine
Muddy R.
Colorado
Moab
Gateway
Gunnison
Iola
Uravan
Milford
Marysvale
Loa
Caineville
Hanksville
La Sal
Steens Mine
Naturita
Beaver
Newhouse Cu.
Junction
Torrey
Ouray
Smith 1826
Henry Mts.
Telluride
Silverton
Iron Mines 1851
Cedar City
Panguich
Circle Cliffs
Escalante R.
Hite
Monticello
Dove Creek
Rico
Bakers Bridge (Gold Mines)
Enterprise
Blanding
Old Spanish Trail 1830
Alton
Bluff
San Juan River
Cortez
Durango
Leeds
Pahreah
Ute Indians
Jepson Spr.
Rockville
St. George
Kanab
Adairville
Johnson
Rainbow Bridge
Navajo Mt.
Carrizo
Shiprock
Aztec
El Vado
Short Cr.
Kaibab Indians
Lees Ferry
Escalante
Farmington
Jacob Lake
Kayenta
Lukachukai
Sonastie
Chilchinbito Spr.
Indians
Chaco Canyon
Navajo
Trumbull
The Gap
North Rim
Supai
Grand Canyon
Cardenas 1540
Escalante 1776
Oraibi
Walpi
(46)
La Ventana
Tovar 1540
Fort Defiance
Gallup
Fort Wingate
Chaves
Little Colorado
Oñate 1604
Oñate 1598
(200)
Estevan killed 1539
Zuni
El Moro
Grants
Laguna
Acomo Pueblo
San Fidel
Williams
Flagstaff
Leupp
Winslow
Colorado R.
Holbrook
Zuni Pueblo
(29)
Coronado (B) 1540
Enchanted Mesa
Sedona
Mormon Lake
Zuni R.
Del Rio
Jerome
Montezuma Well
St. Johns
Prescott
Osborne
Camp Verde
C. Spr.
Quemado
Oñate 1604
Weaver Mt.
Peoples Valley
Octave
Payson
Swansea
Congress
(90)
(51)(92)
(5)
Young
Springerville
Rich Hill
Congress Jct.
(7)
(28)
Butler Well
Wickenburg
Agua Fria R.
Verde R.
(37)
Tonto Cr.
Cibecue
Salt R.
White R.
Ft. Apache
(33)
Alpine
Black R.
Apache Creek
San Francisco R.
(27)(88)(89)
Salome
(16)(40)
Harqua Hala
(32)
Chloride
Mormon Flat
Phoenix
(48)
Superstition Mts.
(34)(35)(50)(51)(57)
Globe
San Carlos
(45)
Miami
Gila
Cu.
Apache
Tempe
Hassayampa
Pinal Mt.
(56)
Hermosa
Cochillo
Hot Springs
Clantons Well
Kino 1691
Gila R.
Geronimo
Morenci
Clifton
Estevanico 1539
Caballo
Gila R.
Florence
Hayden
Ft. Thomas
Aravaipa
Cliff
Gila River
Butterfield Overland Mail
Gila Bend
Casa Grande
(32)
(31)
Ft. Grant
Duncan
Gila R.
Silver City
Apache Camp
Santa Rita (Cu)
San Lorenzo
Battalion
Sentin

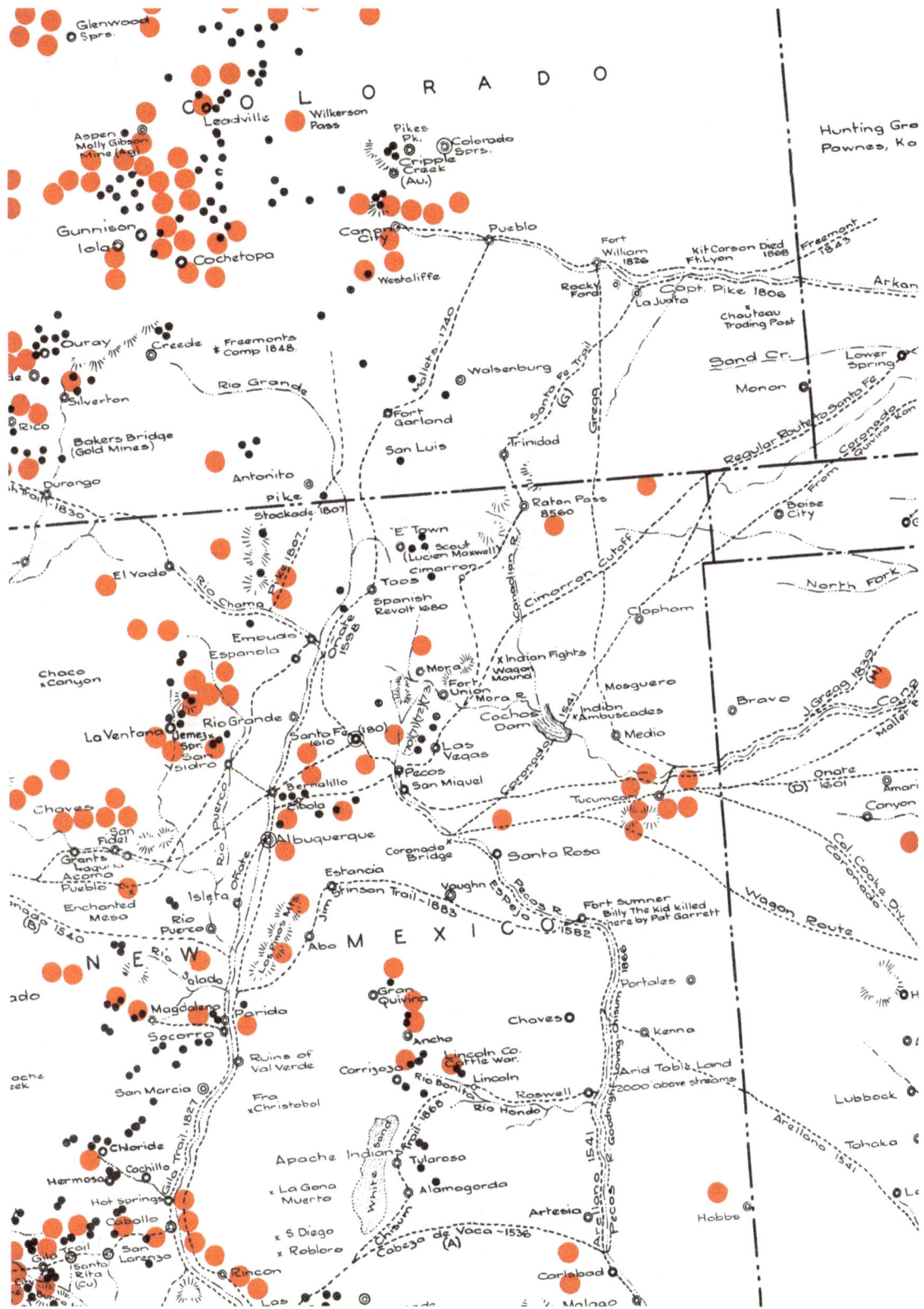

COLORADO

Glenwood Sprs.

Aspen
Molly Gibson Mine (Ag)

Leadville

Wilkerson Pass

Pikes Pk.
Cripple Creek (Au.)

Colorado Sprs.

Hunting Gro
Pawnee, Ko

Gunnison
Iola

Cochetopa

Cañon City

Westcliffe

Pueblo

Fort William 1826

Rocky Ford

La Junta

Kit Carson Died Ft.Lyon 1868

Capt. Pike 1806

Freemont 1843

Arkan

Chouteau Trading Post

Ouray

Creede

Freemonts Comp 1848.

Rio Grande

Walsenburg

Santa Fe Trail (G)

Sand Cr.

Monon

Lower Spring

Silverton

Rico

Bakers Bridge (Gold Mines)

Fort Garland

San Luis

Trinidad

Regular Route to Santa Fe

Coronado Quivira Kon From

Durango

Antonito

Pike Stockade 1807

Raton Pass 8560

Boise City

El Vado

Rio Chama

1827

'E' Town
Scout (Lucien Maxwell)
Taos
Cimarron

Spanish Revolt 1680

Onate 1598

Canadian R.

Cimarron Cutoff

Clapham

North Fork

J. Gregg 1832

Embudo
Espanola

Chaco Canyon

La Ventana
Jemez Spr.
San Ysidro

Rio Grande

Mora
Fort Union
Mora

Cochiti Dam

Coronado 1541

Indian Fights
Wagon Mound

Mosquero

Bravo

Cana

Mallate

Indian Ambuscades

Media

Santa Fe (180)

Las Vegas

Pecos
San Miquel

Tucumc

Onate 1601

Amari

Canyon

Chaves

San Fidel
Grants
Laguna
Acoma Pueblo

Rio Puerco

Cibola

Albuquerque

Bernalillo

Coronado Bridge

Santa Rosa

Pecos R.

Col. Cooke Dix Coronado

Wagon Route

Isleta Otote

Estancia

Rio Puerco

Enchanted Mesa

1540 (B)

Abo

Grinson Trail - 1883

Vaughn Espejo

Fort Sumner
Billy The Kid killed here by Pat Garrett

NEW MEXICO

Salado

Magdalena

Parida

Socorro

Ruins of Val Verde

San Marcia

Fra Christobal

Gran Quivira

Ancho

Corrijoja

Rio Bonito

Lincoln Co. Cattle war.

Lincoln

Chaves

Kenna

Roswell

Rio Hondo

Portales

1866

Arid Table Land
2000 above streams

Lubbock

Arellano Tohaka

1541

Chloride

Hermosa

Cochillo

Hot Springs

Cabollo

San Larenzo

Gila Trail

Santa Rita (Cu)

Rincon

Apache Indian

White Sand

La Gena Muerto

S Diego

Roblero

Cabeza de Voca ~1536

Tularosa

Alamogorda

Artesia

Hobbs

Carlsbad

Malaga

Arellano 1541
Pecos

Gila Trail 1827

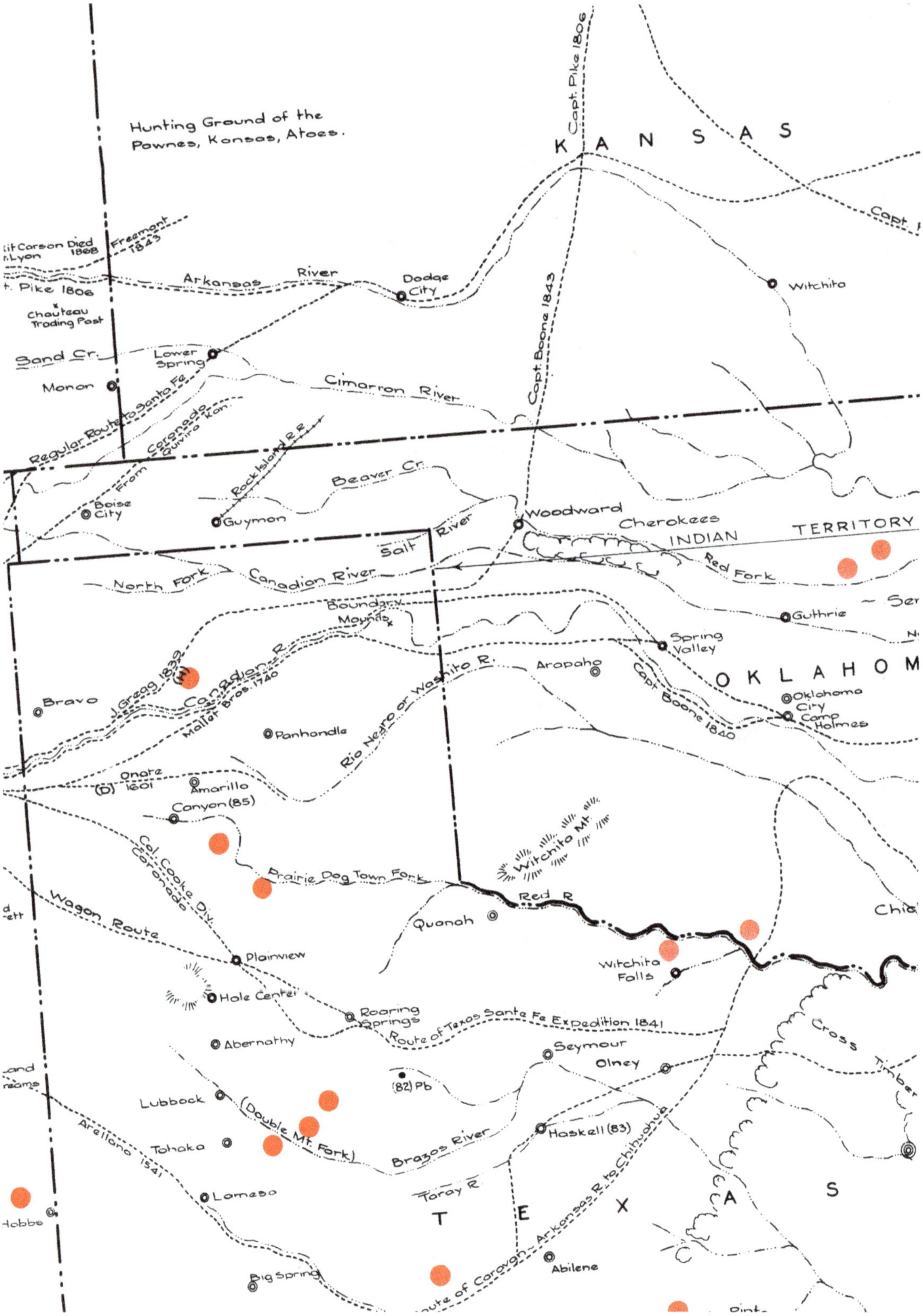

Hunting Ground of the Pownes, Konsos, Atoes.
Capt. Pike 1806
K A N S A S
Capt.
Kit Carson Died Ft. Lyon 1868
Freemont 1843
Arkansas River
Dodge City
Witchita
Chauteau Trading Post
Capt. Boone 1843
Sand Cr.
Lower Spring
Monon
Cimarron River
Regular Route to Santa Fe
Coronado Quivira Kan.
From
Rock Island R.R.
Beaver Cr.
Boise City
Guymon
Woodward
Cherokees
INDIAN TERRITORY
Salt River
Red Fork
North Fork
Canadian River
Guthrie
Boundary Mounds
Spring Valley
J. Gregg 1839
Canadian R.
Mallet Bros. 1740
Rio Negro or Washito R.
Arapaho
O K L A H O M
Capt. Boone 1840
Oklahoma City
Camp Holmes
Bravo
Panhandle
Onate 1601
Amarillo
Canyon (85)
Col. Cooke Div. Coronado
Prairie Dog Town Fork
Witchita Mt.
Red R.
Wagon Route
Quanah
Witchita Falls
Plainview
Hole Center
Roaring Springs
Route of Texas Sante Fe Expedition 1841
Cross
Abernathy
Seymour
Olney
Timber
(82) Pb
Lubbock
(Double Mt. Fork)
Arellano 1541
Tohaka
Brazos River
Haskell (83)
Lomesa
Toray R
Arkansas R. to Chihuahua
T E X A S
Abilene
Big Spring
Route of Coronado

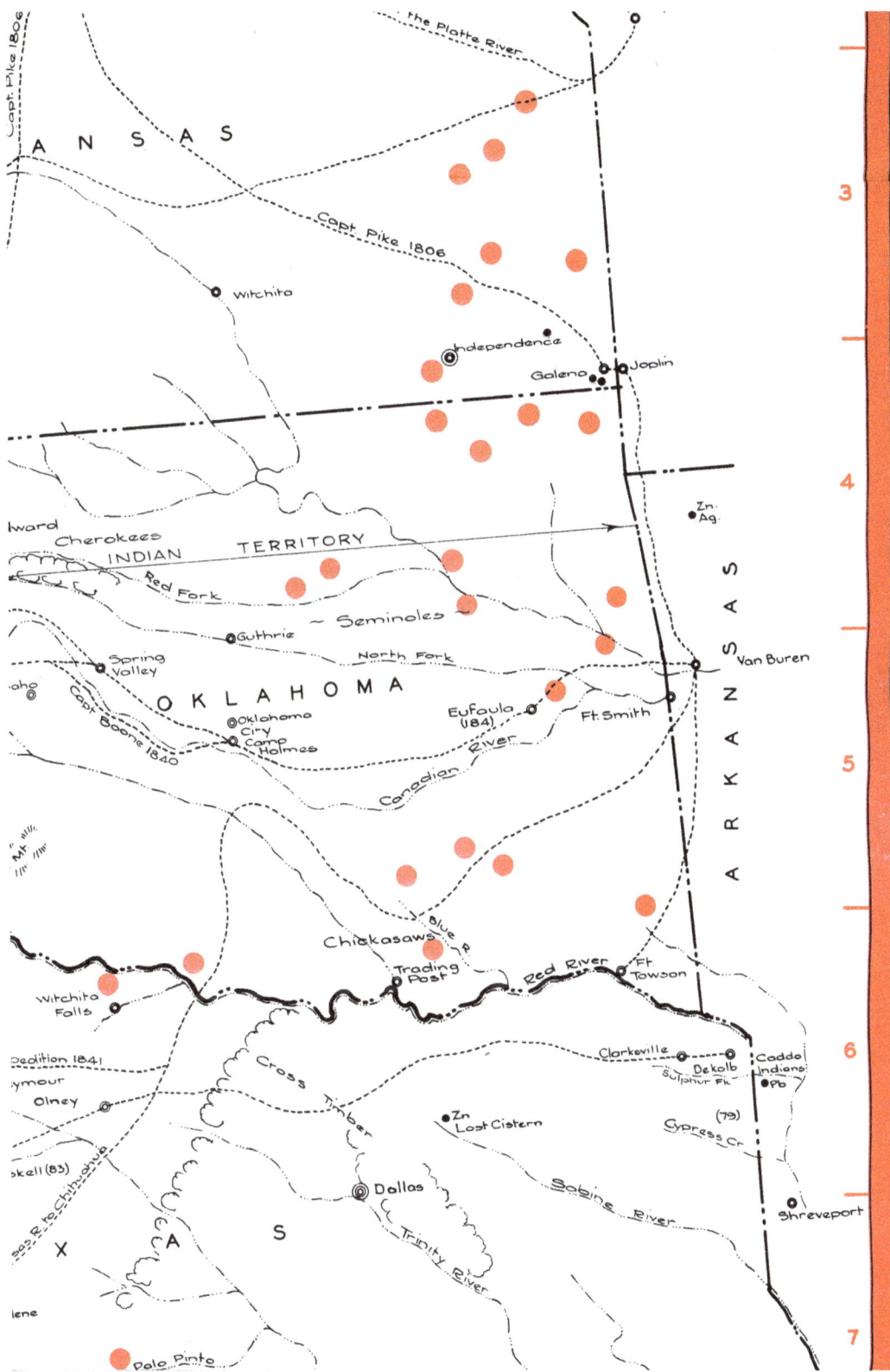

the Platte River

K A N S A S

Capt Pike 1806

• Witchita

•Independence

Galena•• •Joplin

Zn.
•Ag.

Iward
Cherokees   INDIAN   TERRITORY

Red Fork

•Guthrie   ~ Seminoles

North Fork

•Spring
Valley   Van Buren

aho○

O K L A H O M A

Capt Boone 1840

•Oklahoma
City
•Camp
Holmes

Eufaula
(184)   •Ft.Smith

Canadian   River

Mt.

Chickasaws   Blue R.

Trading
Post   Red River   •Ft.
Witchita   Towson
Falls

edition 1841   Clarkeville•
Cross   •Caddo
ymour   Dekolb Indians
•Olney   Sulphur Fk. •Pb
Timber   •Zn   (79)
Lost Cistern   Cypress Cr

skell (83)
os R. to Chihuahua

•Dallas   Sabine   River

T   X   A   S

Trinity River   •Shreveport

ene

•Palo Pinto

URANIUM PROSPECTING

When prospecting for uranium in the Colorado River Plateau Areas —"The Four Corners"— formations will be dealt with which are principally sandstone.

If the search is carried on along the Continental Divide, the same formations that contain silver, copper, vanadium and other minerals should be looked over and vein systems will be studied.

Those areas many miles from either the Colorado River Plateaus or the Continental Divide should be checked around mines that have produced silver, copper, cobalt, iron, lead and vanadium. Check areas showing evidence of hot springs, fluorspar, phosphate, alunite and some of the coal and asphaltic provinces.

Study the mineral areas shown on the map then by the use of the geologic information correlate given areas. That is, a material of a certain geologic age, for instance paleozoic, may be found in New Mexico and also in California, thus affording a chance of finding similar kinds of minerals in the two places.

The geologic study will familarize the prospector with the terms used for a differentiation of the several layers of sedimentary beds of the Colorado Plateau.

These beds are horizontal or tilted except where disturbed by anticlines, folds, faults or by igneous intrusions. Prospect around these disturbed areas. Of the many sedimentary beds that range in age from Paleozoic to Tertiary there are fourteen known formations which contain uranium.

The uranium deposits in the Colorado Plateau are bedded deposits in sandstone, limestone and shales and in general parallel to the bedding.

Geographically these deposits cover a wide area. From Greenriver, Utah, they extend to Holbrook in Arizona, a distance roughly of three hundred miles; from Skull Creek, Colorado to Grants, New Mexico; and from Rifle, Colorado to Leeds in southwestern Utah.

Some of the principal formation types with the kind of ore follows.

1) Entrata — Roscoelite. High Vanadium, low uranium.

(2) Morrison — Yellow Carnotite. Copper, Selenium, silver.

(3) Shinarump Temple Mountain — High uranium low vanadium with selenium, copper, arsenic, cobalt and iron.

(4) Shinarump Conglomerate — High Carnotite.

(5) Chinle — Yellow Carnotite with blue-green copper and silver stains. Leeds, Utah.

(6) Jurassic — Todilto Limestone. Carnotite, uranophane and tuyuyamunite. Grants, New Mexico.

(7) Jurassic Asphaltite deposits — Carnotite associated with vanadium. Found in Utah, California and Arizona.

(8) Miocene — Rosamond — Autunite. California. Generally whether in beds or veins, the uranium ores are irregularly distributed and in relatively small deposits.

Most of the uranium minerals are brightly colored, yellow or green and the rich pitchblende and uraninite ores are of a distinctive black, grayish color, perhaps with copper and iron stains. These ore bodies are recognized in outcrops but this seldom occurs. Careful study of geological and mineralogical conditions together with the use of portable "counters" to test the radioactive ores is in order.

There are many good portable counters on the market which range in price from $25 to $2,000.

LEGEND

● Existing mines or mining districts.

(A) Reference to stories pertaining to trails.

(2) Reference to stories pertaining to lost mines or treasure.

● Area of proven or potential radioactivity

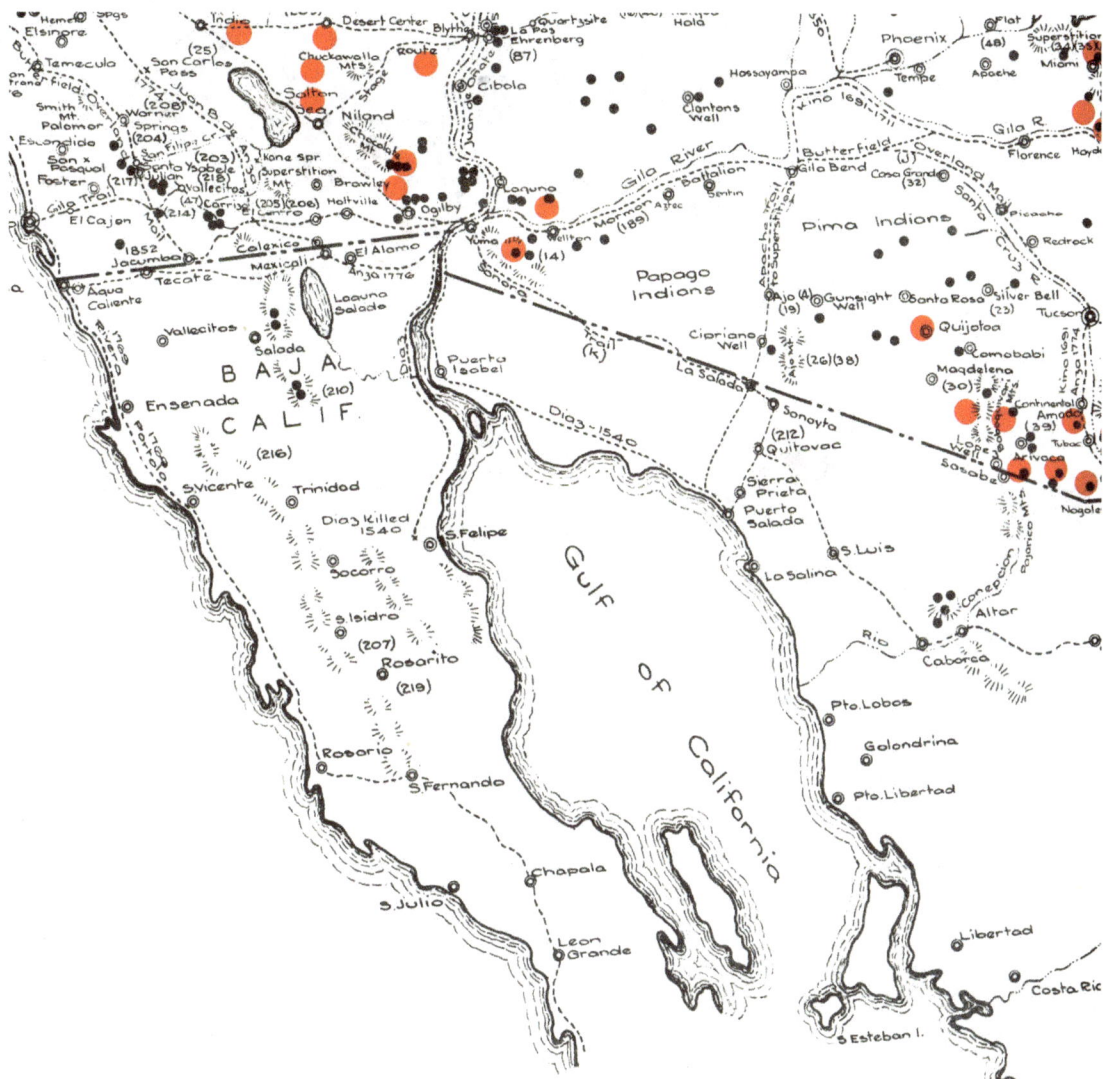

BAJA CALIF.

Gulf of California

Pima Indians

Papago Indians

Phoenix
Tempe
Apache
Miami
Superstition
Florence
Hoyd
Gila R.
Picache
Redrock
Silver Bell
Tucson
Quijotoa
Comobabi
Magdalena
Continental
Amado
Tubac
Kivaca
Sosabe
Nogale
Santa Rosa
Gunsight Well
Ajo
Cipriano Well
Sonoyta
Quitovac
Sierra Prieta
Puerto Salada
S.Luis
La Salina
Conception
Altar
Rio
Caborca
Pto.Lobos
Golondrina
Pto.Libertad
Libertad
Costa Ric
S.Esteban I.

Indio
Desert Center
Blythe
La Paz
Ehrenberg
Quartzite
Hola
Flat
Route
Chuckawolla Mts.
Salton Sea
Niland
Cibola
Hossayampa
Clantons Well
Gila River
Butterfield
Gila Bend
Casa Grande
Overland Mail
Sentin
Aztec
Mormon
Wellton
Yuma
Sonora
Pto.Isobel
La Salada
Diaz 1540
San Carlos Pass
San Jose
Warner Springs
San Felipe
Punta Yaqobe
Superstition Mt.
Browley
Holtville
Ogilby
Laguna
El Centro
Calexico
El Alamo
Mexicali
Ano 1776
Tecate
Laguna Salada
Salada
Ensenada
Vallecitos
S.Vicente
Trinidad
Diaz Killed 1540
S.Felipe
Socorro
S.Isidro
Rosarito
Rosario
S.Fernando
Chapala
S.Julio
Leon Grande
Hemet
Elsinore
Temecula
Smith Mt.
Palomar
Escondido
San Pasqual
Foster
Gila Trail
El Cajon
Jacumba
Agua Caliente

1852

Publisher and Distributor

THOMAS BROS.
Map Publishers Since 1915

LOS ANGELES 39
2560 Glendale Blvd.
NOrmandy 3-9247

OAKLAND 12
337 17th Street
GLencourt 1-6756

SAN FRANCISCO 14
2308 Market Street
Underhill 3-1256

copyright by R. W McALLISTER

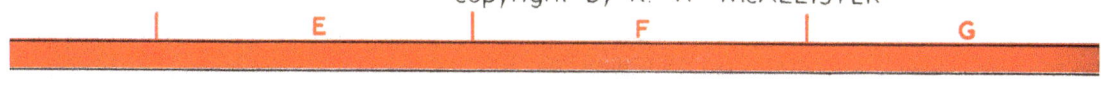

E   F   G

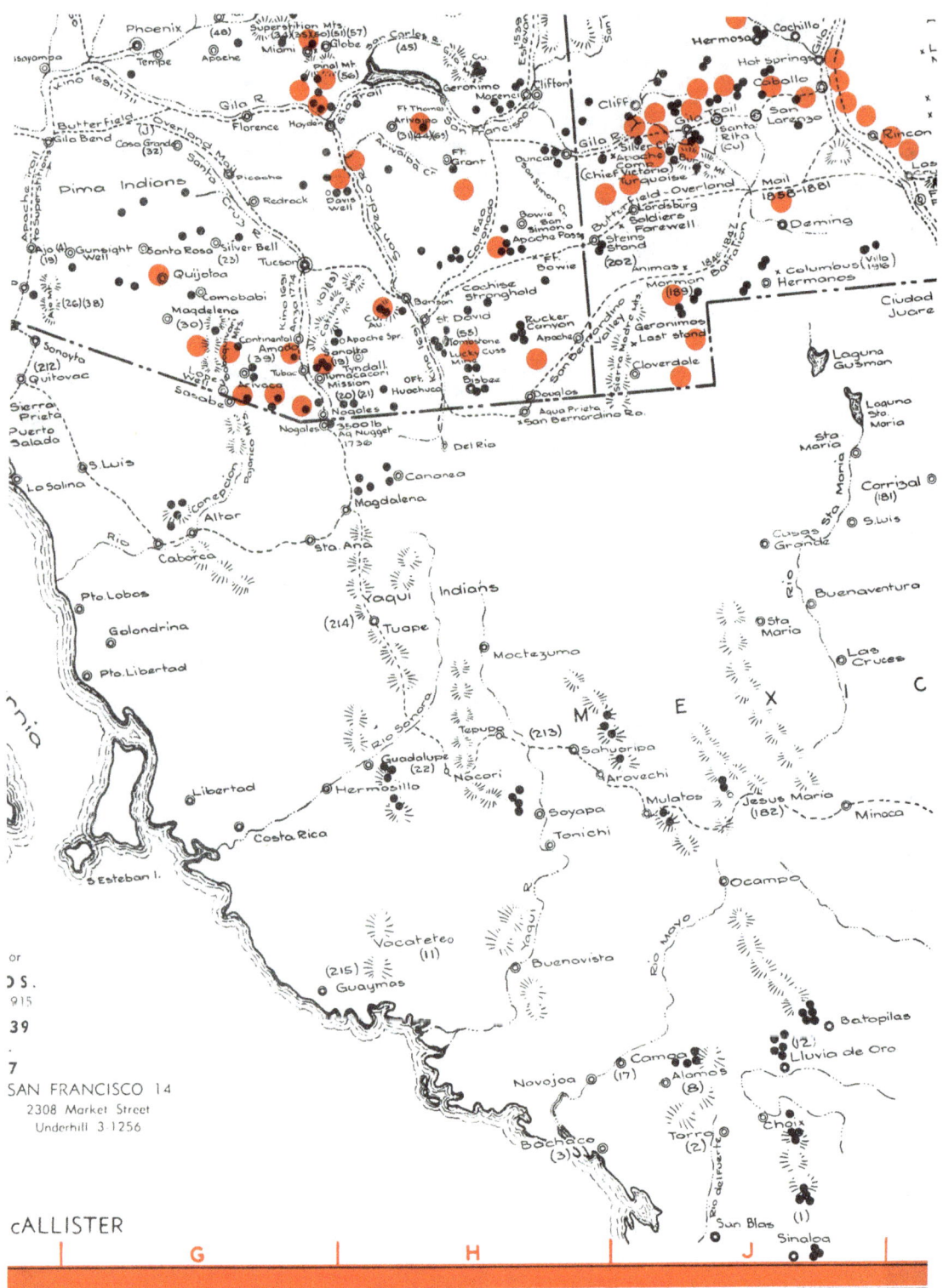

Phoenix · Superstition Mts. · San Carlos · Cachillo · Hermosa · Hot springs · Caballo
Isayampa · Tempe · Apache · Miami · Globe (45) · Clifton · Cliff · Santa Rita · San Lorenzo · Rincon
Pinal Mt. (56) · Gila R. · Geronimo · Morenci · Gila · Silver City · Lordsburg · Los
Butterfield · Overland Mail · Florence · Hayden · Ft. Thomas · San Francisco · Duncan · Apache · Chief Victorio · Turquoise · Columbus
Gila Bend · Casa Grande (32) · Santa · Picacho · Aravaipa Cr. · Ft. Grant · San Simon Cr. · Steins · Deming
Pima Indians · Redrock · Ft. Bowie · Bowie · San Simono · Apache Pass · Soldiers Farewell · Steins Stand · Butterfield - Overland Mail 1858-1881
Ajo (4) · Gunsight Well · Santa Rosa · Silver Bell (23) · Tucson · Animas · Columbus (Villa) (196) · Hermanos
Quijotoa · Comobabi · Davis Well · Ft. Bowie · Mormon · Ciudad Juarez
Magdalena (30) · Continental · Apache Spr. · Cochise Stronghold · Geronimos Last stand
Sonoyta (212) · Amado (39) · Tubac · Tyndall Mission · St. David · Rucker Canyon · Apache · Cloverdale · Laguna Gusman
Quitovac · Sasabe · Nogales · Tumacacori Mission (20)(21) · Tombstone · Lucky Cuss · Bisbee · Douglas · Laguna Sta. Maria
Sierra Prieta · Puerto Salado · S. Luis · Nogales · 3500 lb Ag Nugget 1736 · Del Rio · Agua Prieta · San Bernardino Ro. · Sta. Maria
La Salina · Rio · Cananea · Corral (181)
Concepcion · Altar · Magdalena · Casas Grandes · S. Luis
Pto. Lobos · Caborca · Sta. Ana · Buenaventura
Golondrina · Yaqui Indians · Sta. Maria · Las Cruces
Pto. Libertad · (214) · Tuape · Moctezuma
Rio Sonora · Tepupa · (213) · M E X I C
Libertad · Guadalupe (22) · Nacori · Sahuaripa · Arovechi · Jesus Maria (182) · Minaca
Hermosillo · Soyapa · Mulatos
Costa Rica · Tonichi · Ocampo
S. Esteban I. · Vacateteo (11) · Rio Mayo
(215) · Buenovista · Batopilas
Guaymas · Lluvia de Oro (12)
SAN FRANCISCO 14 · Camoa · Alamos (8) · Choix
2308 Market Street · Novojoa (17)
Underhill 3-1256 · Torro (2) · Sun Blas (1)
Bachaco (3) · Sinaloa

cALLISTER

G   H   J

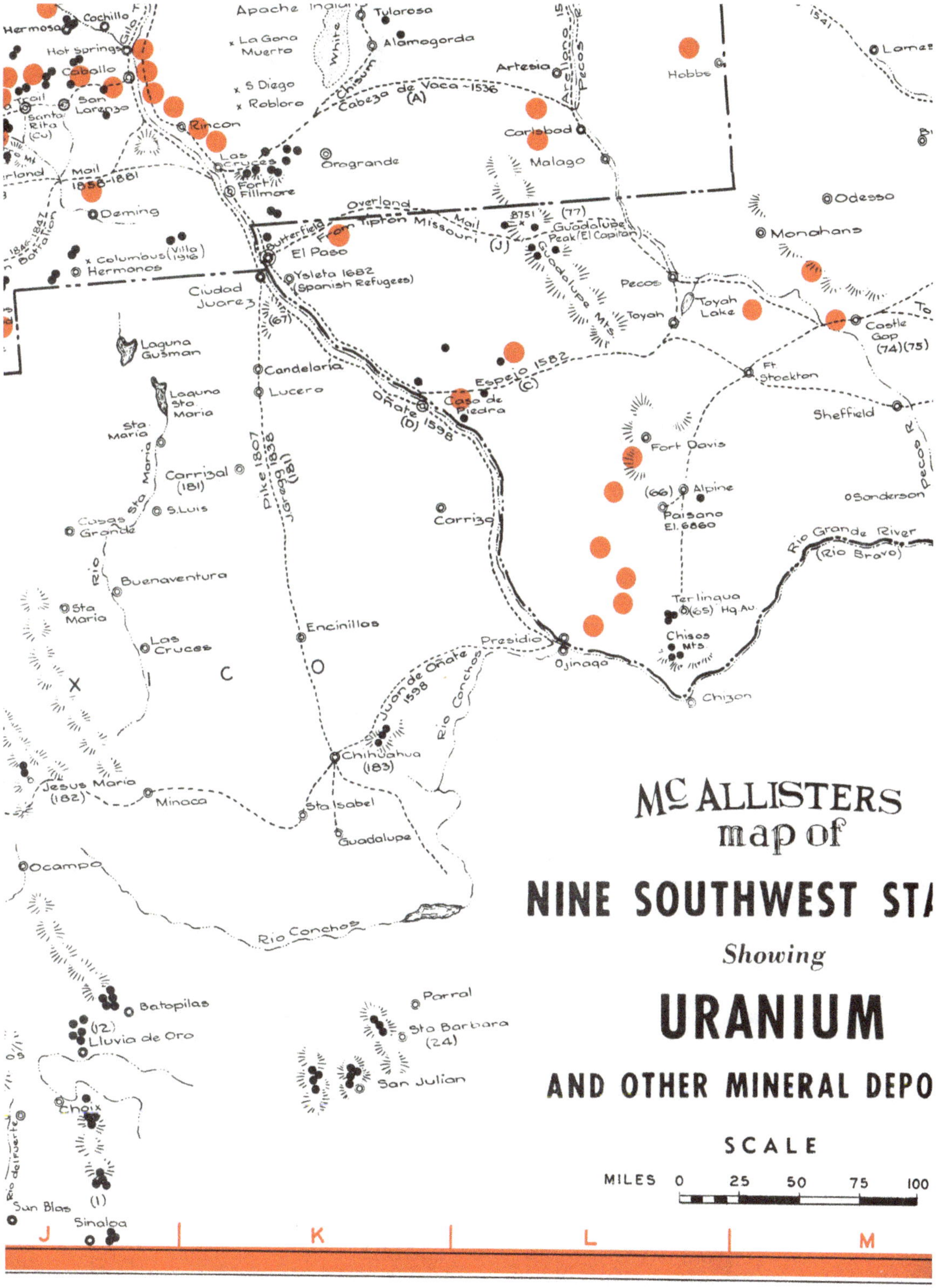

Hermosa Cochillo
x La Gena Muerto
Tularosa
x La Gena Muerto
Alamogorda
Hot Springs
Caballo
x S Diego
Artesia
Hobbs
Lame
Trail
San Lorenzo
x Roboro
Cabeza de Vaca ~1536
(A)
Santa Rita (Cu)
Rincon
Carlsbad
Las Cruces
Mail 1855-1861
Fort Fillmore
Orogrande
Malago
Odesso
Overland
3751
(77)
Deming
Butterfield
From Tipton Missouri
Monahans
1846-1847 Doniphan
Columbus (Villa) 1916
Hermanos
El Paso
(J)
Guadalupe Peak (El Capitan)
Pecos
Toyah
Toyah Lake
To
Ciudad Juarez
Ysleta 1682 (Spanish Refugees)
Guadalupe Mts.
Castle Gap (74)(75)
Laguna Gusman
Candelaria
Onate 1598 (D)
Espejo 1582 (C)
Ft Stockton
Lucero
Casa de Piedra
Sheffield
Laguna Sta Maria
Sta Maria
Carrizal (181)
Fort Davis
Rio Grande River (Rio Bravo)
Cusas Grande
S.Luis
Carrizo
(66)
Alpine
Sanderson
Paisano El. 6860
Buenaventura
Terlingua (65) Hq Au
Sta Maria
Encinillos
Chisos Mts.
Las Cruces
C   O
San Juan de Onate 1598
Presidio
Ojinaga
Chizon
Chihuahua (183)
X   I   C   O
Rio Conchos
Sta Isabel
Guadalupe
Ocampo
Rio Conchos
MC ALLISTERS
map of
NINE SOUTHWEST STA
Showing
URANIUM
AND OTHER MINERAL DEPO
Parral
Batopilas
Sta Barbara (24)
Lluvia de Oro (12)
San Julian
Choix

SCALE

MILES 0   25   50   75   100

Sun Blas (1)
Sinaloa

J   K   L   M

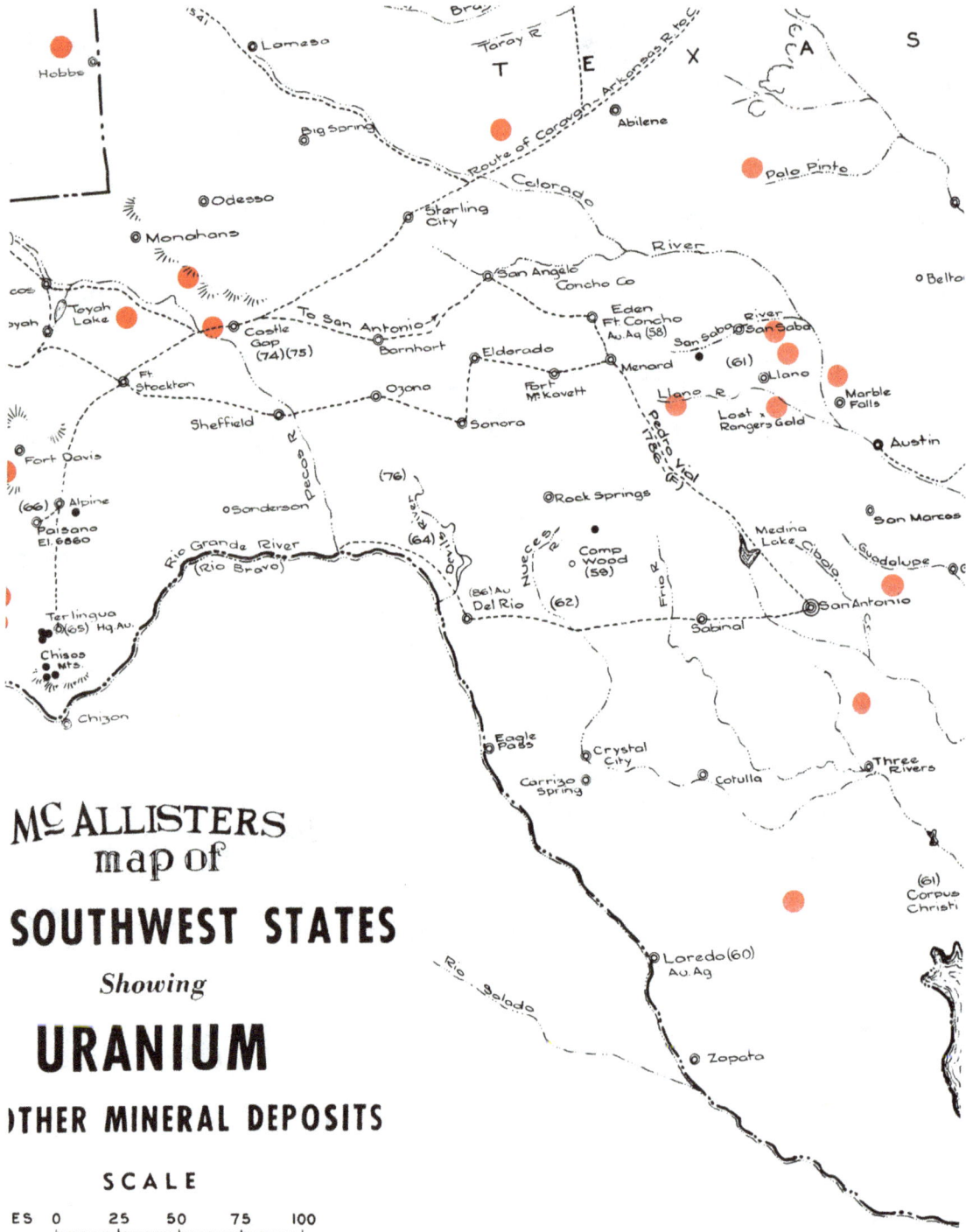

Hobbs

Lamesa

Bray R.
Toray R.

T E X A S
C

Route of Caravan-Arkansas R. to C

Big Spring
Abilene

Palo Pinto

Colorado

Odessa
Starling City

Monahans

cos
San Angelo
Concho Co
River
O Belto

Toyah Lake
Eden
Ft. Concho
Au. Aq. (58)
River
San Saba
San Saba

yah
Castle Gap (74)(75)
To San Antonio
Barnhart
Menard
(61)
Llano
Marble Falls

Ft. Stockton
Elderado
Llano R
Lost x Rangers Gold
Austin

Sheffield
Ojona
Fort McKavett
Pecos R
Pedro Vid
(76)

Fort Davis
Sonora

(66) Alpine
Sanderson
Rock Springs
(64)
San Marcos

Paisano El. 6860
Rio Grande River (Rio Bravo)
Medina Lake Cibolo
Guadalupe

Devils R
Comp Wood (58)
Nueces R
Frio R

Terlingua
(65) Hg Au.
(86) Au
Del Rio
(62)
San Antonio

Chisos Mts
Sabinal

Chizon
Eagle Pass
Crystal City
Cotulla
Three Rivers

Carrizo Spring

(61)
Corpus Christi

McALLISTERS
map of
SOUTHWEST STATES
Showing
URANIUM
OTHER MINERAL DEPOSITS

Loredo (60)
Au. Ag

Rio Salado

Zopata

SCALE

ES 0   25   50   75   100

M   N   O

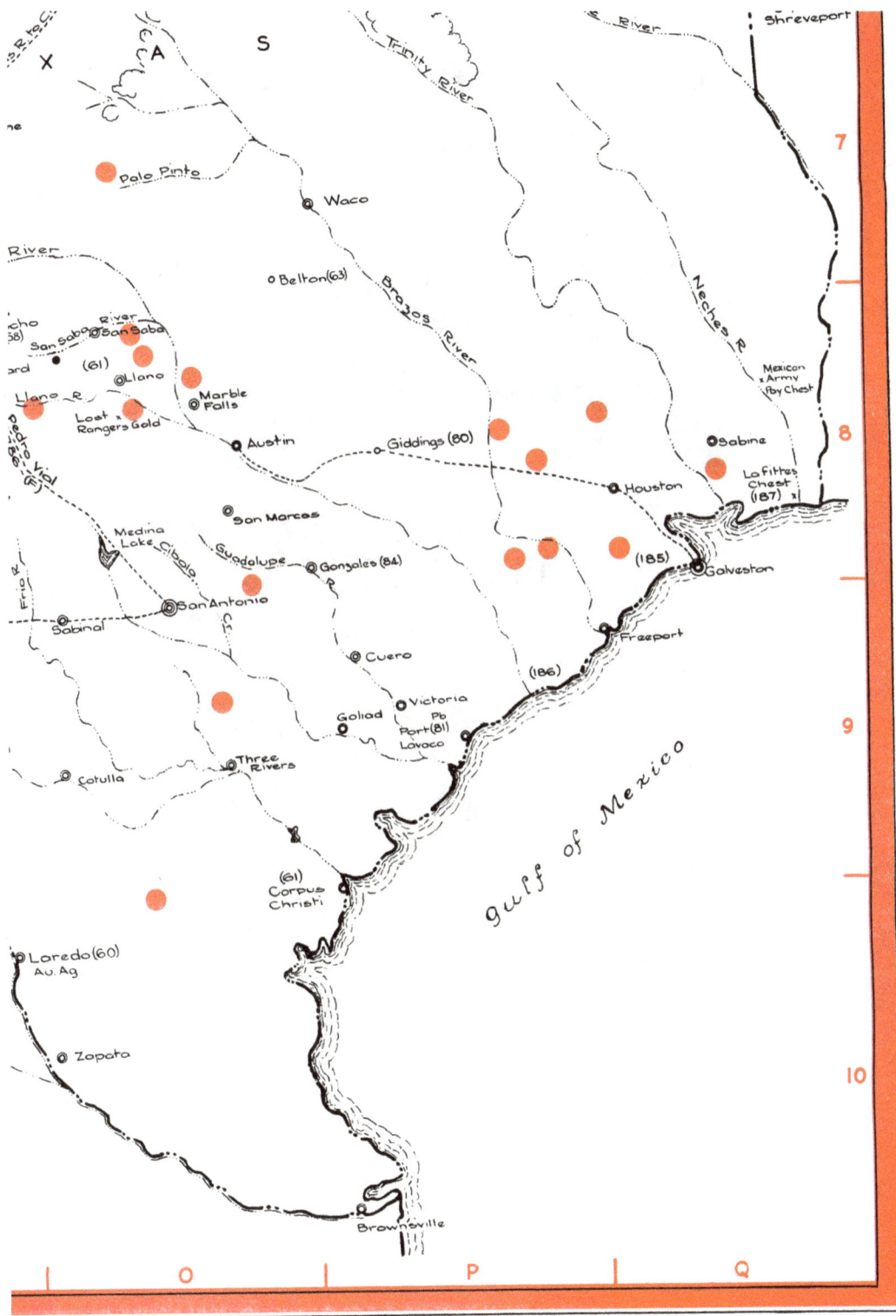

COACHWHIP PUBLICATIONS

COACHWHIPBOOKS.COM

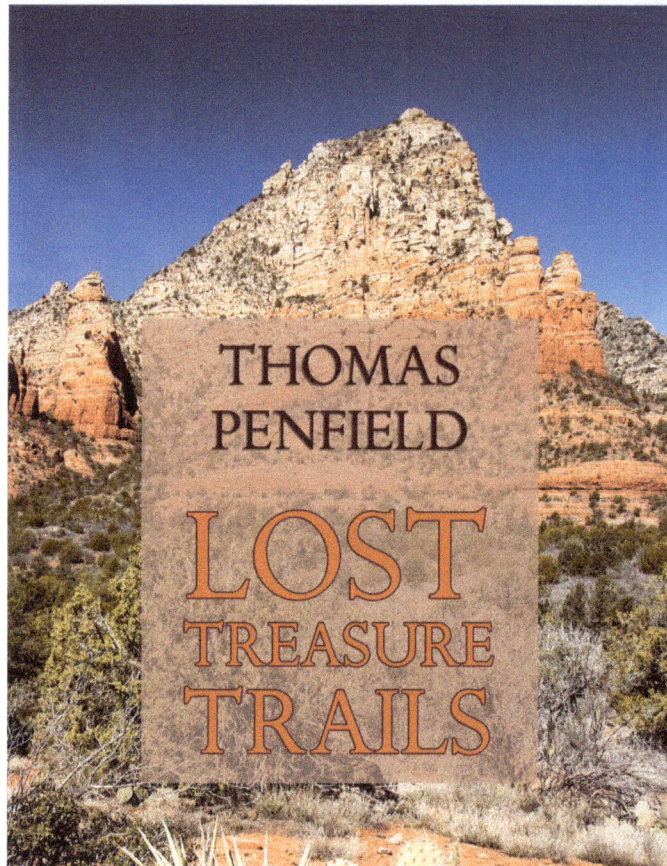

THOMAS
PENFIELD

LOST
TREASURE
TRAILS

Lost Treasure Trails
ISBN 1-61646-218-3

COACHWHIP PUBLICATIONS

ALSO AVAILABLE

Lost Mines of the Old West
ISBN 1-61646-139-X

COACHWHIP PUBLICATIONS

COACHWHIPBOOKS.COM

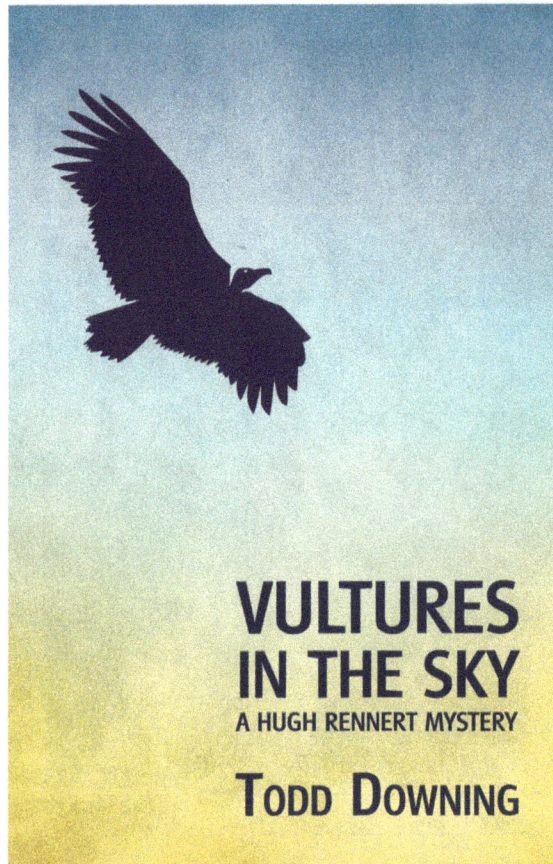

VULTURES IN THE SKY

A HUGH RENNERT MYSTERY

TODD DOWNING

Vultures in the Sky
ISBN 1-61646-149-7

COACHWHIP PUBLICATIONS

ALSO AVAILABLE

PRIMITIVE
AND PIONEER
SPORTS
FOR RECREATION TODAY

Bernard S. Mason

Primitive and Pioneer Sports
ISBN 1-61646-126-8

COACHWHIP PUBLICATIONS

COACHWHIPBOOKS.COM

THE PEACEMAKER
AND ITS RIVALS

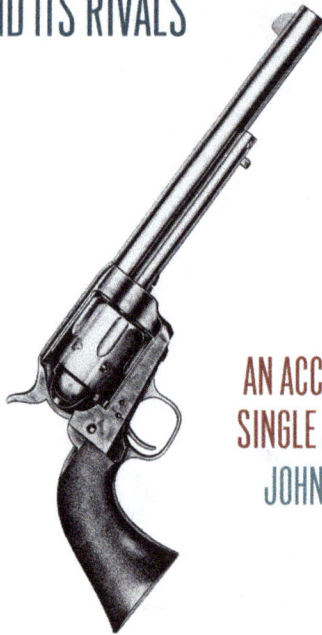

AN ACCOUNT OF THE
SINGLE ACTION COLT
JOHN E. PARSONS

The Peacemaker and Its Rivals
ISBN 1-61646-221-3

COACHWHIP PUBLICATIONS

ALSO AVAILABLE

Missie: The Life and Times of Annie Oakley
ISBN 1-61646-217-5

www.ingramcontent.com/pod-product-compliance
Lightning Source LLC
Chambersburg PA
CBHW062114090426
42741CB00016B/3414